How
HERBS
Healed the World

AND OTHER STORIES OF REMARKABLE PLANTS

How
HERBS
Healed the World

AND OTHER STORIES OF REMARKABLE PLANTS

CONNOR SMITH

greenfinch

Contents

Introduction

Every herb has a story. The dried herbs in the kitchen cupboard, the spice aisle in the supermarket and the potted plant on the windowsill all have complex histories, but we often take their availability for granted. The seasonality of produce is far less of a concern today than it was for ancient humans, who took great care in harvesting their beloved plants from nature; for them, timing was essential, as was knowledge of the edible and functional plants, and of those that were dangerous. As agricultural communities developed about 10,000 years ago, humankind transitioned from the nomadic lifestyle of hunting and gathering to one that involved domesticating plants and animals in an established 'home'.

It is this that has allowed the world's population to grow over the millennia from around 5 million people back then to nearly 8 billion today. Unsurprisingly, this has changed the way we live. Most people in developed countries can now buy herbs, grown across the world, all year round at local shops. We now have the luxury of using herbs every day, from the black pepper (p. 210) that seasons our food, to the meadowsweet (p. 108) in our aspirin, to herbal teas and the sweeteners in fizzy drinks. We may have advanced as a society, but our reliance on herbs and the natural world has not ceased.

People have always used plants for their medicinal, spiritual, edible, ornamental and narcotic value. When we look back in time, we see that humans are inextricably linked with plants and the natural world. The very earliest humans, searching for remedies for diseases and injuries, looked for drugs in nature. As the pursuit of

medicinal plants became more scientific, the possibility of a profitable drug trade made any discovery as much a commercial opportunity as a medicinal one. Herbs have created empires. Trade routes, such as the famous Silk Road (see p. 188), were established for the exchange of prized herbs. People risked their lives on dangerous expeditions by sea and land to places unknown in order to discover new herbs. Some became conquerors. They went to make riches, not to trade, and took both herbs and countries by force. This resulted in the colonization of such pivotal and profitable trade locations as India, Indonesia and South Africa. For herbs, wars were waged, and lives taken or subjected to extreme cruelty and slavery. Powerful companies, such as the Dutch East India Company, used herbs to form monopolies for the rich of the world, creating markets in which a few wealthy countries reigned at the expense of Indigenous communities and poorer nations. These companies caused irreversible damage to the environment through overharvesting, a practice that endangered plants and, in some cases, caused them to be lost forever.

The World Health Organization estimates that some 4 billion people across the globe use herbal medicine as their primary source of medical treatment, and that 80 per cent of people in the 'advanced' nations rely on herbal medicines for some of their healthcare. It is thought that 50,000–70,000 species of plant (out of an estimated 350,000 in the world) have medicinal use, but these are only the ones that are known to science and have been studied. Tens of thousands have not been investigated, or have been investigated only partly, while others have not even been discovered yet. As technology develops even further, we will surely discover more and more about the benefits of plants as medicine and food.

Before we delve into the world of herbs, it is important to understand exactly what a herb is. There are two definitions: the botanical and the cultural. The word 'herb' derives from the Latin *herba*, 'grass': a non-woody plant that begins growing in the spring and dies back in the autumn – a herbaceous plant. Botanically speaking, this excludes trees and shrubs. However, many trees and shrubs are used for herbal healing and in food. Such species as cloves (p. 213; more commonly thought of as a spice), rosemary (p. 30) and St John's wort (p. 103), despite not being herbaceous, are commonly considered to be herbs. This book, then, uses the cultural definition of a herb: a plant, or any part of a plant, that has medicinal, culinary or therapeutic uses.

In *How Herbs Healed the World* I tell the stories of 75 such plants, examining their healing properties, historical uses, folklore and present-day applications, and summarizing scientific research into their uses, benefits and dangers. The herbs I have chosen grow in a range of conditions, from sunny, rocky Mediterranean landscapes to humid tropical rainforests. The selection is also inclusive of a variety of forms, from annuals to long-lived trees, with useful parts that include fruit, buds, leaves and bark. Most herbs have developed a range of adaptations that allow them to survive and grow outside their native range. In those new locations some are grown and valued as ornamental plants, while others make a nuisance of themselves as unwanted weeds. I have aimed to showcase the variety of species that are beneficial to humans, and the range of uses we have discovered or created.

The herbs have been carefully chosen to demonstrate a range of historical applications and, where possible, scientifically proven uses in the present day. I have deliberately excluded some well-known plants, such as Madagascar periwinkle (*Catharanthus roseus*), yew (*Taxus baccata*) and witch hazel (*Hamamelis virginiana*), in favour of lesser-known herbs that are important to Indigenous communities, are regionally important, have a rich history or have become increasingly popular and widespread in recent years.

The herbs are divided into seven chapters to highlight the diversity of their origins and applications: culinary, mystical, narcotics and poisons, healing, indigenous, monastic and exotic. Each entry is accompanied by an illustration, a short description of the herb (including how it grows, its botanical features, and related species or genera) and an in-depth look through time from its traditional uses to its present-day applications. Where relevant, I refer to herbalists who used the plant, the etymology of the botanical or common names, and the groups of people who have used the plant the most.

Through this book, I hope to spread the knowledge of herbs so that we can approach the familiarity our ancestors had with these crucial plants. It is my intention to tell the stories of the plants we grow in our gardens and allotments, the herbs we eat, use to flavour and add spice to our meals, and take as medicine to keep us healthy and improve our overall well-being. I hope the information within these pages instils in the reader a greater appreciation for the natural world and what it gives us.

The information I have included in this book about the use of herbs in traditional remedies and their healing properties is for reference purposes only. This book is not intended to be used as a guide to self-diagnosis or self-treatment and is not a substitute for medical advice. You should consult a qualified healthcare professional on any matters requiring diagnosis or medical attention.

Sage; see p. 29.

CULINARY HERBS

Culinary Herbs

When many of us think of herbs, we think of food. We think of the herb containers in our kitchens or pantries, and the flavours they impart; we think of the potted plants on our windowsills and in our gardens; and we think of the herbal teas we drink for energy, for relaxation and to quench our thirst. Herbs have been used in cooking since humans first made the effort to combine foods. Some flavours have become closely tied to certain countries, regions and cultures: basil (p. 26), for instance, is associated with Italian cuisine, while dill (p. 14) is iconic of Scandinavian cooking.

The details of how herbs are used in cooking have been developed and refined through trial and error over millennia. Herbs are different depending on their site, the climate, light intensity and time of harvesting, and these qualities can change the strength of their taste. The structure of the herb also affects how it is cooked and its ultimate flavour. Soft herbs, such as basil, parsley (p. 21), dill and chives (p. 22), are best added at the end of cooking or used to garnish dishes. Woodier herbs, such as rosemary and thyme (pp. 30, 38), impart their flavour earlier in the cooking process, and are generally not eaten. Drying herbs has long been a popular way of preserving the taste and health benefits of herbs during the months when they are not in season or cannot be grown, although the flavour of fresh herbs is generally preferred.

The word 'spice' is often associated with culinary herbs. A spice is an aromatic or pungent plant used to flavour food. Spices are typically tropical plants and are frequently referred to as 'exotic' species; cumin, black pepper and cloves (pp. 201, 210, 213) are common examples. They are mostly used dried, and primarily to flavour, colour and preserve food. Many herbs are also aromatic – such as coriander seeds – and are therefore commonly referred to as spices. Some people consider only leafy greens to be herbs, making distinctions about what is a herb and what a spice based on the part of the plant that is used. For example, ginger (p. 194) – from an underground rhizome (fleshy root) – and cinnamon (*Cinnamomum verum*) – from bark – are not always understood as herbs. Flavour is another way in which people recognize differences between spices and herbs. Plants with mellow, soothing tastes are often viewed as herbs, while those with warming, spicy flavours tend to be referred to as spices. These terms are loosely formed, easily interchangeable and overlap based on scientific, cultural and culinary use.

Long before the invention of cold storage, aromatic herbs were used to preserve food. This discovery undoubtedly came from early attempts to mask the flavour of decomposing meat. Such herbs as rosemary, thyme and sage (p. 29) were pressed on to the meat, and their antiseptic and antibacterial qualities were revealed when it was preserved. The fact that the flavours of the herbs could penetrate the meat and change its taste was another revelation. Herb preservation later extended to prolonging the shelf life of drinks, particularly with hops (p. 25). It was the ancient Greeks who began to take more notice of flavouring food, investing in new herbs, and experimenting with and adopting flavours from other countries.

This chapter considers common culinary herbs, species familiar from kitchen, garden and food. First I consider where they came from and how each grows in its natural habitat. I then explore the folklore associated with each herb, who used it, and its importance to them, followed by its medicinal properties both historically and in the present day. Finally, I explore the famous dishes each herb appears in – both native recipes and dishes in its adopted home.

These herbs are easy to grow and can be found in farmers' markets, nurseries or garden centres as potted plants or seeds. They can often be grown inside as a seasonal house plant, or outside as a permanent container plant, or simply in the ground. Herbs are easy to harvest; just wash off any dirt and store them in a jar of water, or in the fridge wrapped in damp kitchen paper. They can also be left to dry, then stored in an airtight container for use up to a year later, depending on the species.

Seeds for the stomach

DILL *ANETHUM GRAVEOLENS* (APIACEAE)
COMMON USES SCANDINAVIAN CUISINE • FLATULENCE

The name 'dill' originates from the old Norse word *dylla*, meaning to soothe or lull. This comes from its use under the pillow or hanging above the cot to lull crying babies back to sleep. While dill's original home is believed to be southwestern Asia and India, it is now considered native to lands from the Mediterranean to South Asia. It requires warm summers and high light levels to produce a good yield. India and Pakistan are currently the largest producers of dill, and Hungary of its essential oil.

This tall annual, with its yellow umbrella-shaped inflorescences (flowers), shares a family with such other common culinary plants as carrots, coriander (p. 17) and fennel (p. 18), all of which are characterized by their flat-topped flower heads. What is most commonly called 'dill' is normally the seeds of the plant, while the dried or fresh leaves are called dill weed and have a sharp, spicy flavour. The species name, *graveolens*, from the Latin *gravis* (heavy) and *olens* (smelling), refers to the herb's intense scent, which is attributed to the volatile chemicals carvone in the seeds and myristicin in the foliage. Both have an aniseedy or lemony aroma.

The 1st-century Greek physician and pharmacologist Pedanius Dioscorides believed that dill had many health benefits, and he prescribed it so frequently that for centuries after his death it was known as the 'herb of Dioscorides'. Gladiators in ancient Rome ate dill in the belief that it gave them strength. It has traditionally been used to treat jaundice, headaches, nausea, and stomach and liver problems. When colonists took dill to North America around 300 years ago, it was mixed into an infusion called dill water that Indigenous healers adopted to treat jaundice, scurvy and 'dropsy' (oedema or congestive heart failure).

This last use is comparable with today's gripe water, an alcohol-free drink that contains dill-seed oil, fennel and sometimes ginger (p. 194). It is used to relieve the symptoms of colic, and is mostly administered to children. Herbalists also use dill as a tea to treat stomach aches, colds and coughs, and to increase the production of breast milk, while the seeds can be chewed to prevent bad breath. When added to a warm bath, dill oil can help to expel urinary tract infections.

Dill is important in central European cuisine, often used with butter to garnish boiled potatoes and to season fish. It is a quintessential ingredient of Scandinavian dishes, and is considered a natural pairing with fatty fish, such as salmon, herring and mackerel. Traditionally, fresh fish may be placed in a dish and *gravad* (buried) in a dry cure of salt, sugar and spices for a few days, then served with dill, butter and rye bread, or with Swedish *Färskpotatissallad* (potato salad). In Sweden, meat, fish and vegetables are cured, pickled or fermented to preserve them for the long winter. *Västeråsgurkor* – gherkins mixed with dill, mustard seeds, peppercorns and a few other ingredients, including *ättika* (a strong vinegar) – is one example.

Dill is not only used as a flavouring, but also has a long history as a natural preservative for pickling foods.

Coriander produces spherical seeds that are light brown in colour and have a slight lemony flavour.

'Love it or hate it' garnish

CORIANDER *CORIANDRUM SATIVUM* **(APIACEAE)**
COMMON USES SALADS · CURRIES

This fragrant and flavourful herb has for centuries been used in traditional dishes in countries from Peru to India. It is known in Britain as coriander (an adaptation of the French name, *coriandre*) and in the United States as cilantro (from the Spanish name for the herb). Other commonly used names include Chinese parsley and Greek parsley. *Coriandrum* comes from the Greek word *koris*, which has several meanings; none is particularly pretty, but the most relevant is probably 'stink bug', referring to the rather earthy smell of the foliage.

All parts of coriander are edible, even the flowers. This is a 'love it or hate it' plant, and some say that depends on our genes; a bit like the burning sensation some people get when eating raw brassicas, coriander leaves for some taste unpleasantly soapy. But the plant's popularity is saved by the rather cleaner aromatic quality of the seeds, particularly once dried, toasted and ground. Those seeds that are grown in India are large and slightly bitter, whereas those from northern Africa are sweeter and more aromatic. Coriander seeds grown in Europe have more lemony notes, and are used mainly for essential oils.

A native of the Mediterranean and the coastal Middle East, coriander has been cultivated for more than 3,000 years. It is mentioned in ancient Hindu texts and Egyptian medical papyri, and even in the Bible. In Exodus 16.31 manna – the bread given by God to the people of Moses – is referred to as 'white, like coriander seed'.

Seeds have been found at archaeological digs in ancient Roman sites, as at Silchester in Hampshire, southern England, proving that this tender annual herb was probably introduced to and grown in the British Isles by Roman settlers. Although it was not widely grown in Britain after that, it does reappear in literature throughout the Elizabethan period, being used medicinally to settle the stomach and often to mask the taste of less flavoursome herbal medicines. We can assume that cultivation continued, since there is reference in the gardener and herbalist Maud Grieve's *A Modern Herbal* (1931) to coriander being grown extensively in the county of Essex for flavouring gin. In Germany, meanwhile, coriander seed became a key flavouring for sausages.

World merchandise tables show that India leads the market for growing and exporting coriander seed to satisfy the massive demand for it in the West. This earthy, citrussy seed is frequently combined with ground cumin seeds as the basis of fragrant curries of all kinds.

Coriander is a familiar annual crop in the vegetable patch, happiest in warm, well-drained soil. Some cultivars are slow to bolt (run to seed), putting on strong-flavoured leafy growth, while others quickly produce seed for harvesting. If you leave the seed to drop to the ground it will germinate easily and give a good supply all season.

Prometheus's gift

FENNEL *FOENICULUM VULGARE* **(APIACEAE)**
COMMON USES INDIGESTION • SPICE

Fennel is a statuesque plant that can grow up to 2.5m (8¼ft) tall. It produces light, feathery leaves and an array of yellow, umbrella-like flowers. The word *foeniculum* is Latin for 'hay' and refers to the feathery foliage. It is also the root of the English word 'fecund', meaning something that is fruitful, fertile and abundant – just like the common (*vulgare*) fennel. The plant is cultivated for its aromatic foliage and tasty seeds. Large amounts of the chemical anethole make the aroma similar to that of anise (*Pimpinella anisum*). The plant is variable in growth, and different cultivars have been selected for certain characteristics. Some flower more enthusiastically, and are valuable forms for gardens, while others are more aromatic or have a better flavour. Florence fennel (*F. vulgare* var. *azoricum*) has an inflated base that appears swollen. Its leaves have a sweeter flavour and are more aromatic than those of the typical species, and the bulb itself is used as a vegetable.

The ancient physicians Hippocrates (4th century BCE) and Dioscorides shared the opinion that fennel increased milk production in nursing women. The Romans are said to have chewed fennel seeds after meals to cleanse their palates, reduce flatulence and stimulate digestion. Indeed, the chemical chloride, which is present in this herb, helps the body to produce hydrochloric acid, a key component in gastric acid, which aids the breakdown of food.

According to ancient Roman folklore, the consumption of fennel seeds triggered the shedding of a serpent's skin. Fennel also features in Greek mythology. The Greeks believed that Prometheus – the creator of humankind, who stole fire from the gods and gave it to humans – kept the stolen smouldering ember in the hollow stalk of a fennel plant. Interestingly, the Greek common name for fennel is *marathon*, from the fact that the plant grows abundantly around the city that gave its name to the famous race now run in many cities across the world.

The fennel bulb is frequently grilled, sautéed or roasted with other herbs and lemon. It is generally cooked in quarters, after the tops, fibrous outer leaves and bottom section of the root are removed. In Italy it is also braised in milk, sugar and spices, to make a traditional dish known as *finocchi al latte*. The fresh leaves, meanwhile, are finely chopped and added to seafood and pasta dishes, and the traditional dish *tullore* is made with chestnuts and cooked fennel leaves. Fennel is a very popular herb in Crete, where it is used to flavour meat, fish and vegetables. In Germany, a spice mix called *Brotgewürz*, containing fennel seeds, aniseed and coriander seeds, is used to make bread. Fennel seeds are also commonly added – whether whole or ground to a fine powder – to German sauces, soups and stews.

Puritans in the 16th and 17th centuries reportedly ate fennel seeds
during church services to suppress their hunger.

Never pick parsley in the wild; it looks very similar to the deadly
water hemlock, poison hemlock and fool's parsley.

Championed by Charlemagne

PARSLEY *PETROSELINUM CRISPUM* (APIACEAE)
COMMON USES GARNISH • BREATH FRESHENER

Parsley comes from lands around the Mediterranean, specifically Algeria, Tunisia and Morocco, and possibly the Balkan countries, but is also now common in other areas, and in similar climates around the world. The common name 'parsley' comes from the old English *petersilie*, which resembles other European names, among them the Dutch *peterselie*, German *Petersilie*, French *persil* and Spanish *perejil*. Parsley is a small herb that is normally annual or biennial, meaning that it dies after producing seeds in its first or second year. There are two main varieties: curly-leaf and flat-leaf. It is important not to mistake parsley for 'fool's parsley' (*Aethusa cynapium*), which is poisonous. The strong resemblance between fool's parsley and flat-leaf parsley has made the curly variety increasingly popular.

Given parsley's present ubiquity, it is surprising that it features so little in ancient texts. The first known mention of it is attributed to the Greeks, who considered the herb to be sacred, laying it on tombs and using it to adorn the winners of athletic contests. The 8th-century Roman emperor Charlemagne was a great advocate of parsley and is credited with the spread of the plant's popularity.

Parsley has traditionally been used for its diuretic properties, and to treat autoimmune and inflammatory illnesses. A clinical trial conducted in 2017 by researchers from Gonabad University in Iran found that the topical use of the herb made a notable difference in individuals suffering from melasma, a skin condition that affects pigmentation. In the cosmetics industry, an essential oil produced from parsley is used in creams and soaps. However, certain compounds present in the essential oil can lead to increased menstrual flow, and it has been known to result in miscarriage in pregnant women.

Parsley is very easy to grow in the home or garden, and is usually ready to harvest between 70 and 90 days after being planted. This is typically done after watering and in the morning, to ensure the stems don't wilt. Once picked, it is best to trim the ends of the stems, wrap the parsley in damp kitchen paper and put it into a closed plastic bag in the fridge, where it can last for up to three weeks. However, dried parsley can retain its flavour for up to a year and is often favoured over fresh for cooking.

Parsley can be found in cuisines throughout Europe, the Middle East, the Americas and Asia, used both as garnish and for flavour. The curly-leaf variety tends to be used as a garnish, and the flat in cooking. In the traditional English dish called pie, mash and liquor, the 'liquor' is parsley sauce. In Lebanese cooking – specifically the country's national dish, *tabbouleh* – parsley is chopped finely and mixed with bulgur wheat and chopped tomato to accompany falafel. In Iran, parsley and other herbs are combined to make the ancient and still very popular stew *ghormeh sabzi*.

Chopped chives

CHIVES *ALLIUM SCHOENOPRASUM* (AMARYLLIDACEAE)
COMMON USE GARNISH

The English word 'chive' comes from the Middle English *cheve* and the Old French *cive*. The species name, *schoenoprasum*, derives from the Greek *schoinos* (rush) and *prason* (leek). The etymology of *allium* is unknown, although it may come from the ancient Greek for onion, *aglis* or *aleo*. However, it is well known that the genus *Allium* contains a range of notable culinary herbs, among them onion, garlic (p. 120), leek and shallot. It belongs to a family that contains such ornamental bulbs as *Galanthus* (snowdrops), *Narcissus* (daffodils) and *Agapanthus*. Chives – sometimes referred to as European chives – are closely related to Chinese or garlic chives (*A. tuberosum*). However, the former has a slightly weaker flavour and tends to be used as a garnish rather than being incorporated into dishes, as the latter frequently is.

Although chives have spread across temperate countries in the northern hemisphere, Siberia is believed to be their native area. The tale goes that the Siberians gave chives – supposedly an aphrodisiac – to Alexander the Great of Macedonia before his marriage to Princess Roxana in 327 BCE. According to Romanian lore, gypsies hung chives from the ceiling and bed to ward off evil spirits. In East Asia, there are some records of chives being used as remedies for flu and maladies of the lungs. Members of the genus *Allium* in general are known to reduce blood pressure and have antimicrobial properties, and their high potassium content makes them helpful in treating kidney stones.

Recent studies have shown that phenolic compounds present in allium flowers are antiproliferative, so allium can reduce the risk of gastric and prostate cancer.

Chives are milder in taste than garlic and onion. Unlike with those herbs, it is not the bulbs that are eaten, but the leaves and occasionally the flowers. With a long culinary history on the Continent, chives are a common ingredient in dishes from many European countries, and are frequently used as a garnish for scrambled eggs and omelettes, or added to potato salad, seafood, chicken and cheese dishes. In Sweden, they are an important part of the traditional *gräddfil* (similar to sour cream), which is used to make such iconic dishes as *toast skagen* (shrimps on toast). In France, chives feature in the cold potato soup *vichyssoise*. They are often paired in Poland with *tvorog* cheese and in Germany with *Quark*. In Slovenia, chives are used in a *potica* (traditional bread) called *želševka*.

When buying chives, it is important to select bunches that are a vibrant green colour and have a light oniony fragrance. If they are wilted or slimy, or give off a strong odour, they are no longer fresh. They will last longest in a vase of water, or wrapped in damp kitchen paper and stored in the fridge. They are also easily grown as a pot plant on the windowsill or outside, for fresh trimmings at any time.

Chives are very easy to grow and can be used as a garnish,
as cut flowers, or as an ornamental plant in the garden.

Hops can be grown from seed or propagated by root cuttings in the spring or autumn.
They are very vigorous, and a mature plant can be 8m (25ft) high.

Behind the brew

HOP *HUMULUS LUPULUS* (CANNABACEAE)
COMMON USES BREWING · PRESERVATIVE

This climbing plant is dioecious, meaning that each example is either male or female. For hops' most famous application – brewing – only the female plants are used, since the male flowers impart an unpleasant taste to the beer. The genus name, *Humulus*, is understood to come from Low German, while the species *lupulus* relates – along with the common flower (or, to some, weed) *Lupinus* (lupin) and the disease lupus – to the Latin word *lupus* (wolf). All are known for their rapid spread and negative impact. Lupins increase the nitrogen content of soil, and can then outcompete the native plants, which cannot grow so well in the changed soil, while lupus attacks the body's immune system. Similarly, the hop has a reputation for colonizing land and shading out other plants. This has led to it being classed as an invasive plant in some countries. The origin of the common name, 'hop', is uncertain, but it may come from the Anglo-Saxon word *hoppan* (to climb), no doubt because the plant is a vigorous and rapid climber.

When exactly hops first began to be cultivated has been subject to much speculation. Brewing is thought to have begun as early as 8,000 years ago, but it probably did not involve the use of hops until the 1st century CE. Indeed, some believe that hops were not involved until as late as the 8th century. Their use almost certainly began in central Europe, and many assert that it was in Germany, which has a rich history of brewing; others believe it was French monks who first developed the technique of brewing with hops.

At a time when clean drinking water was not commonly accessible, beer was considered a safer alternative. Ancient beers were much lower in alcohol than today's brews, typically about 2–3 per cent, as opposed to 4–10 per cent. Hops were initially used to preserve the beer and prevent it from going bad. Historically, hops have been used to make tea and food as well as beer. The fresh shoots were cooked or pickled and added to bread to act as a preservative. Hops were also used to make a tonic that acted as a diuretic and a treatment for stomach problems. Mixed with poppy seeds, hops were used to treat inflammation, swelling and bruises.

Today, an estimated 97 per cent of all commercially produced hops are grown for the brewing industry. Of that total, Washington, Oregon and Idaho in the United States, and the Hallertau region of Germany, account for about 75 per cent of production. The number of small breweries has increased since the 1990s, leading to the creation of a wide range of new flavours and variations of beer, involving different varieties and combinations of hops. Belgium has the richest diversity of beers per capita (about 1,500 from some 400 breweries) of any country in the world.

The king of herbs

BASIL *OCIMUM BASILICUM* (**LAMIACEAE**)
COMMON USE ITALIAN CUISINE

Despite its status as an iconic herb of Italian cuisine, basil is native to tropical and subtropical Asia. It is well known for its scent and oil, the volume of which – as well as the taste and composition of the herb – depends on the climate in which it is grown. This variable herb is often annual in cultivation in temperate climates. Because of its variability, many cultivars have been selected; among them are the Italian variety *O. basilicum* 'Genovese Gigante', which is most commonly found in pesto.

It is thought that an archaic Middle Eastern or Asiatic language was the root of the Greek for basil, *ókimon*, from which the genus name derives. This herb was introduced to the ancient Greeks around 500 BCE, and the name 'basil' comes from the Greek *basilikon* (royal) and *basileus* (king), hence the plant's historical common name 'king of herbs'. *Basilicum* may also bring to mind the mythical lizard-like beast called the basilisk (little king). According to the Roman author Pliny the Elder (writing in the 1st century CE), the crest on a basilisk's head resembled a crown. Cementing the relationship between *basilicum* and the mythical creature, basil was believed to be an antidote to basilisk venom.

In Dioscorides' classic medicinal text *De materia medica* (1st century CE), basil is noted as an antidote for scorpion stings. Still, the ancient Greeks and Egyptians had a largely negative opinion of the plant, possibly owing to its use in embalming. It was thus often seen as a symbol of mourning and loss. As the holy plant of the goddess Tulasi (or Vrinda) – symbolic of wifehood and motherhood – basil is regarded as one of the holiest plants in the Hindu religion. Hindus were buried with basil, which was believed to help them pass into the paradise of Svarga Loka. In India, basil is used in Ayurvedic medicine as an adaptogen, helping the body to respond to stress, anxiety and fatigue. In various parts of the world it is made into a tea for treating nausea and dysentery. Recalling the basilisk venom and the scorpion sting, basil essential oil is used to treat snake bites and wasp stings. In Jewish folklore, it was said to give strength during fasting. Basil has also been used as a cancer treatment in traditional Chinese medicine. The herb is now valued for a range of antiviral, antifungal, antioxidant and anti-inflammatory properties.

Of course, basil's reputation rests mostly on it being a culinary ingredient, and it is particularly popular in the Mediterranean. In Italian cuisine, as well as being a crucial ingredient in pesto, it is often added fresh as a finishing touch to pasta dishes and pizzas. Similarly, it features as a garnish for such Indian dishes as dal and roti, and also in Thai dishes, such as the popular *pad kra pao gai* (basil chicken). In Iranian cuisine, fresh basil is often used in salads, among them *sabzi khordan*, a combination of fresh herbs and raw vegetables served as an accompaniment. Basil seeds can be used as a substitute for chia seed in soups, bread and smoothies.

In Haiti, basil is mixed with water and sprinkled around
a property in the belief that it will ward off evil spirits.

Sage was thought to heal all illnesses and health problems. Some 10th-century
physicians believed it even offered a way to attain immortality.

Savoury stuffing

SAGE *SALVIA OFFICINALIS* (**LAMIACEAE**)
COMMON USES COSMETICS • MEAT DISHES

Sage is a small, aromatic herb belonging to the largest genus in the mint and dead-nettle family. Given that there are more than 1,000 species in the genus, it is unsurprising that the common name 'sage' can refer to any one of many different species. *Salvia officinalis* is native to Italy and the western regions of the Balkan Peninsula, and monks and Roman soldiers expanded its natural range. The name comes from the Latin *salvus* (safe, healed), while *officinalis* simply means that it is used for medicine.

Indeed, the renowned American clinician and expert in herbal medicine Varro Tyler wrote that 'every sickness known to humanity will be listed as being cured by sage.' It was regarded as a sacred herb in ancient Egypt and Greece, being used in Egypt to improve fertility and in Greece to treat ulcers and sores, and as a coagulant (blood-clotting agent). The leaves were used dried or fresh to preserve meat and fish, and tea made with sage leaves was drunk for sore throats and coughs. Some even believed that sage could improve memory and brain power, and sprigs of the herb were placed beneath pillows to repel bad dreams. In the Middle Ages, the plant was used to treat heartburn and bloating.

When it was taken by explorers and traders to Asia and South America, sage gained a role in local folk medicine for inflammation, gout, rheumatism and seizures. In the 16th century the Dutch introduced it to China, where it quickly became a beloved plant – to such an extent that tea was traded for sage at a ratio of three to one. Sage was introduced to North America in the 19th century by colonists, and was adopted by Indigenous healers to treat insomnia, measles and seasickness.

In the genus *Salvia*, the species *S. officinalis* produces the most essential oils. The volatile components in these oils vary according to climate and growing conditions, as well as the plant's age and genetic make-up. In the cosmetics industry, sage essential oils are commonly used in perfumes, deodorant and soaps for their astringent properties, meaning that they reduce sweating and tighten the skin. Sage's range of medical uses is still being investigated. Herbalists now recommend it for cuts, insect bites and even depression, and studies are considering its use as a cancer treatment, since some of its essential oils show promise in the fight against colon cancer. A study in 2010 showed that extracts from sage leaves reduced metastasis – the spread of cancer cells throughout the body – which is the cause of 90 per cent of cancer deaths.

Sage leaves have a light, peppery taste that makes them popular for use in rich dishes. Unlike more delicate herbs, sage is frequently added early in the cooking process. It balances rich pasta sauces in Italian cooking, and is often added to sausage fillings in France. In North America, sage is the quintessential herb in Thanksgiving stuffing and turkey dishes.

Nature's preservative

ROSEMARY *SALVIA ROSMARINUS* (LAMIACEAE)
COMMON USE PRESERVATIVE

Not long ago, rosemary belonged to its own genus and was named *Rosmarinus officinalis*. However, DNA analysis from 2015 onwards has led to its reclassification into the very large genus *Salvia*. While there is no question that the genera were similar, this change came as a shock to some of the botanical community. Many were surprised to see its iconic name, *Rosmarinus*, demoted to that of a species. The word and its derivatives have been well known for centuries: *rosmarin* in Danish, German and Swedish, *romarin* in French, and *romero* in Spanish, all meaning 'rose of the sea', for its habit of growing on dry, rocky soil in maritime conditions.

Rosemary is a strongly aromatic evergreen shrub that, in time (and it is long-lived), produces a gnarled trunk with flaky brown bark. The green leaves are narrow and the woody stems square in section, bearing pale purple flowers typical of its family. Rosemary has become a very common ornamental plant in many Mediterranean climates, where it can flower for months at a time. It is surprisingly tolerant of cold, especially when it is kept dry, and – as for many woody plants of Mediterranean hillsides and cliffs – the biggest problem is damp, which causes the roots and stems to rot. It is commonly found in supermarkets and also used in flower arrangements. Growing it from seed can often result in poor germination, so cuttings are the preferred method of propagation.

The historical use of rosemary is documented from about 5000 BCE. Branches of this fragrant shrub were placed in the tombs of Egyptian pharaohs to help them pass into the afterlife. In ancient Greece, rosemary was symbolic of love and death; it was used at both weddings and funerals to signify the eternal bond of love and marriage, and branches of rosemary were burned in temples to honour Aphrodite, the goddess of love. Students put rosemary in their hair to improve their memory, a belief that continues in rosemary's use as a symbol of remembrance in many Western countries; as Ophelia famously says to Laertes in Shakespeare's *Hamlet*: 'There's rosemary, that's for remembrance.' The herb was also considered sacred, and was believed to cure various illnesses and ailments. The Romans added it to baths to relieve muscle pain, and it was taken orally to relieve gastrointestinal inflammation and similar disorders. The Romans considered it powerful medicine and were responsible for the spread of the herb outside the Mediterranean.

Rosemary is of particular significance in Hungary, although the details of exactly why are hazy. In the early 13th century Árpád-házi Szent Erzsébet (Elizabeth of Hungary) seems to have become very unwell (although some sources say that she just wanted to look more youthful). Whatever the case, she was cured with rosemary-infused wine applied to her skin. Rosemary wine was subsequently called 'Queen of Hungary's water' and was used to treat skin problems and gout. Later, other

Despite coming from the Mediterranean, rosemary is surprisingly tolerant of cold.
If grown in full sun with good drainage, it can survive to -20°C (-4°F).

Rosemary is a classic household herb, having been hung at doorways to ward off thieves, and from bedposts to protect the sleeper from evil spirits.

herbs were added, and its name was altered to 'Hungary water'.

In Europe in the Middle Ages, the use of rosemary broadened. Like other fragrant herbs, it was popular for strewing on the floors of churches and homes, where it was trampled to release its scent, to repel insects and other pests with its strong fragrance, and for its antifungal and antibacterial qualities. Similarly, rosemary was left in the cots of newborn babies to protect them from evil forces. Rosemary – rich in essential oils, and still regularly used in massage oils, soaps, fragrances, skincare products and shampoo – was first used as a perfume in 1330. It was still considered symbolic of love at this time. Tapping an individual on the shoulder with a flowering rosemary stem was said to make them fall instantly in love. It was present in churches during weddings, and features in bridal bouquets to this day.

In 16th-century England, rosemary became a symbol of femininity and was closely associated with a female-run household. Some men removed rosemary from their homes in anger, believing that their authority was being disputed. During the Great Plague of 1665, rosemary was carried in pouches hung around the neck to fight infectious vapours while travelling, and French nurses during World War II burned it in hospitals as an antiseptic incense.

In the present day the applications of rosemary are numerous. They include aromatherapy, cosmetics and medicine, as well as food production and preservation. Its health benefits are many and various, from improving memory and reducing nervousness and tension, to easing digestion and improving blood circulation and liver function. The plant is frequently used by Indigenous peoples in North and Central America; in Mexico and Guatemala, for example, it is used to treat respiratory problems, skin infections and flatulence. Indeed, according to an article in the *Journal of Ethnopharmacology* in 2019, countries from every inhabited continent have historical and present-day medical uses for rosemary.

The same can be said of this versatile herb's impact on the culinary world. Rosemary has a unique flavour. It is a robust herb that can last well at high temperatures and over a long cooking time, unlike other, softer herbs – such as basil and oregano (pp. 26, 37) – which must be added late in the cooking process, or they will wilt and lose their flavour. It is most commonly used with meat. This habit originates in one of rosemary's original uses, as a preservative, when the crushed leaves would be wrapped around pieces of meat, lending the food a herby aroma and keeping it fresh. Rosemary is currently being investigated for its potential as a food additive to preserve flavour, taste or appearance.

To harvest rosemary, remove some of the new, less woody shoots. Rinse them well to remove any dirt or debris, and dry with kitchen paper. The whole stem (stalk and leaves) can be added to a stew or roast, or the leaves can be removed and finely chopped. The leaves can also be sprinkled on top of Italian-style bread, such as focaccia, mixed with butter or added to potatoes. Additionally, the stems can be used to infuse olive oil with a delicious flavour. Rosemary is also a crucial ingredient of some alcoholic drinks, among them the French herbal liqueur Bénédictine.

Minty fresh

PEPPERMINT *MENTHA × PIPERITA* (LAMIACEAE)
COMMON USE ANTISEPTIC

Peppermint is a naturally occurring hybrid between two closely related species, believed to be *Mentha spicata* (spearmint) and *M. aquatica* (water mint). Unlike some other species of mint, peppermint is best grown in a shady, well-watered spot; without enough water, the leaves tend to flop in warm weather. Mints are well known for spreading very quickly and by runners – in fact, they are now largely classed as invasive species – so it is best to confine them to containers to prevent them from taking over the garden.

According to Greek mythology, Mentha was a beautiful nymph who caught the eye of Hades (the god of the underworld) and became his mistress. They were discovered by Persephone, Hades' wife, who viciously trampled Mentha to the ground. Hades could not bring her back to life but did bring her spirit back in the form of the fragrant earth-hugging plant that now bears her name. The ancient Greeks and Egyptians documented their use of mint, but it is not clear which species they are discussing. The use of peppermint specifically is recorded in Europe from the mid-18th century onwards.

Peppermint oil has many different uses today, including in cosmetics, medicinal and pharmaceutical products, and food. It is applied to areas of inflammation to relieve swelling, and is used in aromatherapy for its antiseptic, antispasmodic and decongestant qualities. It can also be found in mouthwashes and toothpastes for its refreshing taste and fragrance.

In food and cooking, peppermint is primarily known for adding a refreshing flavour to tea, coffee and hot chocolate, as well as chewing gum, desserts and sweets. In the Maghreb (the Arabic countries in northwestern Africa), mint tea is served at most social gatherings, with a generous amount of sugar. The tea varies throughout the region; it is typically sweeter in the north, and in some areas wormwood and lemon verbena (pp. 74, 136) are added. In these hot, dry countries, mint tea is considered more hydrating than cold water.

Morocco dominates production, being responsible for 83 per cent of the world's peppermint. A study in 2017 of the chemical composition of peppermint grown there indicates why. As with all herbs, the essential oils in peppermint vary according to climate, growing season and light intensity. The study found that Moroccan-grown peppermint has a high menthol content, while Algerian peppermint has more trans-carveol, and peppermint grown in Iran a much lower concentration of chemicals overall. The richer peppermint is in menthol, the better its antimicrobial activity, making it more desirable for use in toothpastes and mouthwashes.

Being a hybrid, peppermint tends to grow more vigorously than the true species.

Like all related species, the true oregano needs light, well-drained soil and plenty of sun.

The true oregano

OREGANO *ORIGANUM VULGARE* **(LAMIACEAE)**
COMMON USES ITALIAN CUISINE • INDIGESTION • PRESERVATIVE

The word 'oregano' comes from the Latin *origanus* and Greek *oreiganon*, from *oros* (mountain) and *ganos* (brightness). Many species of this herb are mountain dwellers, as was the Greek mountain nymph Oreias. *Vulgare*, on the other hand, simply means common, and oregano is found throughout the Mediterranean region and has naturalized more widely in the northern hemisphere. It is an aromatic species of variable height, although often short; in dry, harsh conditions it is only 5–10cm (2–4in) high, but in rich soils and at low elevations it can reach 1m (3ft). It produces hundreds of small, pale purple flowers in the summer. These can be cut back, encouraging more flowers and resulting in a show until mid-autumn.

Oregano is much more challenging taxonomically than it is to grow. About 40 different herb species in four different families have been given the common name 'oregano'; for example, *Coleus amboinicus* is called 'Cuban oregano' (as well as 'Mexican mint' and 'Spanish thyme', despite originating in the Arabian Peninsula and Western Asia). Although oregano belongs to the same family as mint and thyme, it is not closely related to them and belongs to a completely different genus.

Origanum vulgare is the species of oregano that is most commonly sold and used. Its medicinal properties were first recorded in ancient Greece and Rome, where it was used to relieve muscle aches and skin sores. Traditional tinctures were thought to reduce the symptoms of asthma, as well as cramping and indigestion. Oregano was also used to preserve meats and fish. In 2006 fragments of olives and oregano were discovered in two 2,400-year-old *amphorae* (ceramic jars) in the depths of the Mediterranean Sea. The survival of these fragments for centuries on the sea floor is a testament to oregano's high antioxidant content.

Oregano contains chemicals called carvacrol and thymol, which have excellent antibacterial qualities. Thymol is used in commercial mouthwashes, and at home, oregano tea – often made with fresh lemon juice – is consumed to ease digestion and cure coughs and colds. Studies of the antioxidant qualities of oregano have shown that the plant may be helpful in combating cardiovascular problems, inflammation and ailments to do with blood glucose. In 1997 the American Herbal Products Association reported no known health concerns over the use of oregano.

Oregano has an earthy flavour with a light bitterness. Unlike many herbs, it is normally used dried because the oils in the fresh herb can be overpowering; it keeps very well in the storecupboard. As well as its ubiquitous use in such Italian dishes as pizza and pasta, it is used to flavour olive oil and to make vinaigrettes, and is common in Turkish and Greek dishes. Mexican cooking often features 'oregano', too, but that is more often *Lippia origanoides*, an unrelated plant in the family Verbenaceae.

Nature's disinfectant

THYME *THYMUS VULGARIS* (LAMIACEAE)
COMMON USES PRESERVATIVE • ANTISEPTIC

Thyme is yet another aromatic herb in the mint and dead-nettle family. Many species of thyme are earth-hugging plants – tough, woody and fragrant – adapted to dry conditions. The small purplish flowers make it an attractive plant for use in the garden, and a must-have for cooking. Thyme is a common sight in Mediterranean landscapes, and it is so familiar in the Spanish landscape that it gave its name (*tomillo*) to a particular area: Los Tomillares in southern Spain. It is native to southwestern Europe but very well adapted to various environments and climates, ranging from hot central Europe to its adopted homes of Cuba, Algeria and the South Island of New Zealand.

Some believe that the word *thymus* comes from the Greek *thyo* (perfume), while others believe it comes from *thymos*, which refers to courage and strength. 'Thymo' was also the name given to the closely related herb savory (*Satureja* spp.). The potential for confusion worsens when it comes to the common name 'thyme', which is generally given to all members of the genus and to some unrelated species, too. This has led to errors in the literature and a lack of clarity as to which species is being referred to. Spanish *origanum* oil, for instance – actually a species of thyme – comes from *Thymus capitatus*.

In ancient Egypt, thyme was used during the embalming process, to which its disinfectant and preserving properties would have made it particularly suited. It is also recorded in the Ebers Papyrus, an Egyptian herbal that dates from about 1550 BCE – making it one of the oldest known medical books – for use as a medication to reduce snoring. It was even thought to help the sleeper have positive dreams. It is doubtful whether *T. vulgaris* was present in Egypt at that time, however; it is more likely that the ancient Egyptians used *T. capitatus*, which is native to the area and has similar properties.

Thyme was used as a strewing herb in ancient Greece to discourage pests and insects, and as a preservative, its foliage chopped and spread over meat to keep it fresh. Similarly, the herb was sprinkled over sacrificial animals to make them palatable as offerings to the gods. Greek soldiers would bathe in thyme-scented water to gain courage before battle, and it was burned in large bathhouses to give the fighters strength and valour. The Romans burned thyme to fend off snakes and other venomous creatures, and used it to aid digestion, to cure colds and to treat intestinal worms. They also took thyme to other countries, including France and England, and monks later helped to spread the species further through the cultivation of monastic gardens.

In the Middle Ages the herb once again became a symbol of courage, and sprigs were sewn into knights' clothing in an attempt to grant them good fortune. Thyme was stuffed into pillows to help people relax, and to ease depression and anxiety. It continued to be used as a strewing herb, and was found both underfoot and

Thyme has historically been rubbed into wounds to clean them before bandaging.

Thyme was a fashionable herb in the past, and among the ancient Greeks the phrase 'to smell of thyme' described someone stylish and elegant.

hung in bunches in medieval homes. The 12th-century German herbalist and abbess Hildegard of Bingen recommended it for skin disorders, and by the end of the Middle Ages its uses were being recorded by English herbalists. In the 16th century John Gerard recommended it for sciatica and leprosy, while in the following century Nicholas Culpeper stated that it could cure headaches, aid during childbirth, help shortness of breath and rid an individual of nightmares.

In the late 17th century thyme oil became available to treat a range of illnesses, including coughs and other respiratory problems. The German chemist Caspar Neumann managed to extract its active component in 1719, and this was later named 'thymol' by the French chemist Étienne Alexandre Lallemand. For the next 200 years, thymol was used primarily as an antiseptic, but it was also highly regarded for its ability to treat headaches, menstrual cramps and indigestion. Given that most thymol was produced in Germany, the beginning of World War I meant a great shortage of antiseptics for the rest of Europe, forcing the Allies to import large quantities of soya oil to use instead. In the years after World War II the use of thymol and thyme as a disinfectant decreased, being replaced by stronger synthetic disinfectants.

Today, thymol is still used as an antiseptic. It is an important ingredient in mouthwashes because it cleans the mouth, prevents the build-up of bacteria and can reduce inflammation, thus keeping the mouth healthy. For the same reasons, it is also used to treat sore throats, usually by being made into thyme honey and added to hot water, or by being infused directly into tea.

Thyme's long folk history of dealing with illnesses is now being backed up by conventional science. In 2021 researchers from Zanjan University of Medical Sciences, Iran, prescribed thyme essential oil for patients who had contracted COVID-19. Those individuals who used it (in addition to their prescribed medication) experienced significantly shorter infections and more manageable symptoms.

The mild flavour of this herb has made it popular in many different cuisines. *T. vulgaris* is commonly used dried. The stems are harvested and gathered into bunches to air-dry, which is best done in a warm, dark room with good airflow and low humidity. The leaves can then be removed easily and stored for later use. Different types of thyme have been selected or bred for their different flavours. For example, *T. × citriodorus* (lemon thyme) is popular in chicken dishes.

Thyme's historical use as a preservative means that it is still frequently paired with meat. Lamb, thyme, butter and garlic is a common combination, and in Middle Eastern cooking thyme features in a spice mixture for meat. The herb was introduced to France by Benedictine monks, who incorporated it into sauces, soups and stews. Such iconic French dishes as coq au vin, confit de canard and beef bourguignon almost always contain thyme.

Lady's mantle; see p. 53.

MYSTICAL HERBS

Mystical Herbs

There is, and has always been, a mystical and magical aspect to herbaceous plants. Seeds are sown and the seedlings emerge, grow leaves and flower, before dying back, only to resprout the following spring. Our ancestors named many of their gods after the natural world, and every continent has had its gods of nature. Harvest gods could yield a crop to feed the people – if they were pleased; if not, they could leave you starving. Sun gods and rain gods could decide if the herbs that fed and healed would grow. Strict rules regarding the collection of plants were followed to please the gods and spirits. The religious elements of herbs have a meaningful and long history; many Indigenous peoples of the world have spiritual connections to herbs.

Beyond the deification of the natural world, people have turned to stories and folklore to explain the wonders of plants and their apparently supernatural abilities. From there it was a short step to spread fear by claiming that plants could be used by demonic forces to make spells and enchantments. Witchcraft was believed to be rife in Europe between 1300 and the 1650s, and there were countless witch hunts that resulted in the execution of women across the continent. Some believe this was linked to the Black Death – bubonic plague – which wiped out half of Europe's population in the 14th century. They contend that a desire to eliminate the spread of the disease resulted in the massive death toll of these 'witches'. Others, among them the historian Lucy Worsley, suggest that these deaths occurred because of a conflict between herbal medicine, practised by 'witches', and secular medicine, an almost entirely male-dominated sphere.

Whatever the reason for this persecution, it changed the perception of herbal medicine. It could be said that the belief that herbal medicine is not truly medicine – that it is a method of healing that has lost its value when confronted by 'modern' science – stemmed from the witch hunts, which killed prominent female herbalists and damned the reputations of those who survived. Knowledge of herbalism consequently dwindled and advances in natural healing slowed as those practitioners who remained went into hiding in order to survive. There are no references to female herbalists in this period because all their works were destroyed or lost. It was only in the 18th century – after the witch hunts had ended – that herbalism experienced a resurgence.

Many of the herbs that were believed to be the most magical before the witch hunts are now classed as common weeds. Borage, hollyhock and agrimony (pp. 58, 61, 65), for example, are adaptable species and have spread far and wide. They are often overlooked, and the casual passer-by is entirely ignorant of their mystical histories. They were used to cast hexes, protective spells and love charms, and for healing. Superstition meant that herbal goods must be purchased using silver. Similarly, amulets made of silver or herbs were made to ward off evil spirits or spells.

This chapter considers the mystical origins of ten herbs – the so-called witches' plants. These are the herbs that were believed to be useful in spells, to be connected to the devil, or to be cursed. The mandrake plant (p. 46), for example, was reported to scream, killing anyone who heard it. Hemlock (p. 62) and other poisonous herbs were supposedly used in spells and enchantments. Contrastingly, agrimony was understood to stop evil and demonic forces, and was symbolic of purity. Angelica (p. 66) – as its name suggests – was related to the divine, at a time when angels were invoked as protection from harm in the form of plagues and evil magic.

Here I investigate the magical qualities of these herbs, from their use by witches and enchanters to conjure hexes, to their function for those people who used them to break spells and counter demonic forces. While examining their history and the lore and rumours attached to these herbs, I reveal their methods of harvesting, their uses and their deadliness. I also look at the people who believed the rumours, as well as those who still use the herbs today.

The screaming mandrake

MANDRAKE *MANDRAGORA OFFICINARUM* (SOLANACEAE)
COMMON USES ANAESTHETIC • WITCHCRAFT • PAINKILLER

Mandragora is a genus from Asia and the Mediterranean, not to be confused with the North American 'mandrake' (*Podophyllum peltatum*), which is unrelated. *M. officinarum*, from the nightshade family (which also includes tomato, potato and aubergine/eggplant), is a herbaceous plant that produces a very thick, long taproot. It also has an intriguingly sinister reputation. Its use as a herb is surprising, given its high levels of alkaloids (organic compounds), particularly tropane alkaloids, which make it poisonous and hallucinogenic. Too high a dose results in delirium, coma and possibly even death.

The myths that have grown up around mandrake may have originated in efforts to safeguard the species. Mandrake has always been regarded as a primary ingredient in magical potions, and some lore suggests that it was a key part of a broom's flying ointment. Given its medicinal and hallucinogenic potential, huge quantities were removed from the wild, so perhaps witches and sorcerers spread the rumours to dissuade people from removing the plants. Others who had a stake in the continued survival of the plant were the *rhizotomoi* (professional root diggers), who harvested plants for their medicinal use and sold them at market. Whatever its origin, the story behind mandrake told of a demon that, when unearthed from beneath the plant's green rosette of leaves, would harm those who disturbed it by emitting a scream that was said to be fatal, or at least to cause madness.

According to the 1st-century Romano-Jewish historian Flavius Josephus, the moon had to be up if one were to dig out the mandrake plant safely. To do so, he said, three circles must be drawn around the plant with a sword, then prayers had to be said and a dog (often black, for superstitious reasons) tied to the plant. The dog would be bribed to lurch forwards, tearing the plant from the ground and eliciting the mandrake's ear-bleeding shriek. Such stories were retold in Byzantine times by St Theophanes the Confessor in the 8th century and in medieval times by the 13th-century herbalist Bartholomaeus Anglicus. They were not questioned until the publication of the *Grete Herball* in the 16th century, and the idea was echoed by Shakespeare in *Romeo and Juliet* (IV:iii): 'with loathsome smells,/ And shrieks like mandrakes torn out of the earth,/ That living mortals, hearing them, run mad'. In 1597 the English herbalist John Gerard claimed that he had 'replanted many mandrakes' and lived to tell the tale. Still, the legendary screeching of the mandrake plant survived in the public consciousness. Students in the first book in J K Rowling's Harry Potter series (1997) are described wearing earmuffs to repot the screaming mandrake.

The mandrake seems always to have possessed a negative image. It was said in ancient times to grow on the cliffs of Greece, beside men who had been hanged, and on the graves of those who had taken their own lives. Modern studies have proven

Mandrake belongs to the family that also contains tomatoes,
and its fruit is strikingly similar to unripe tomatoes.

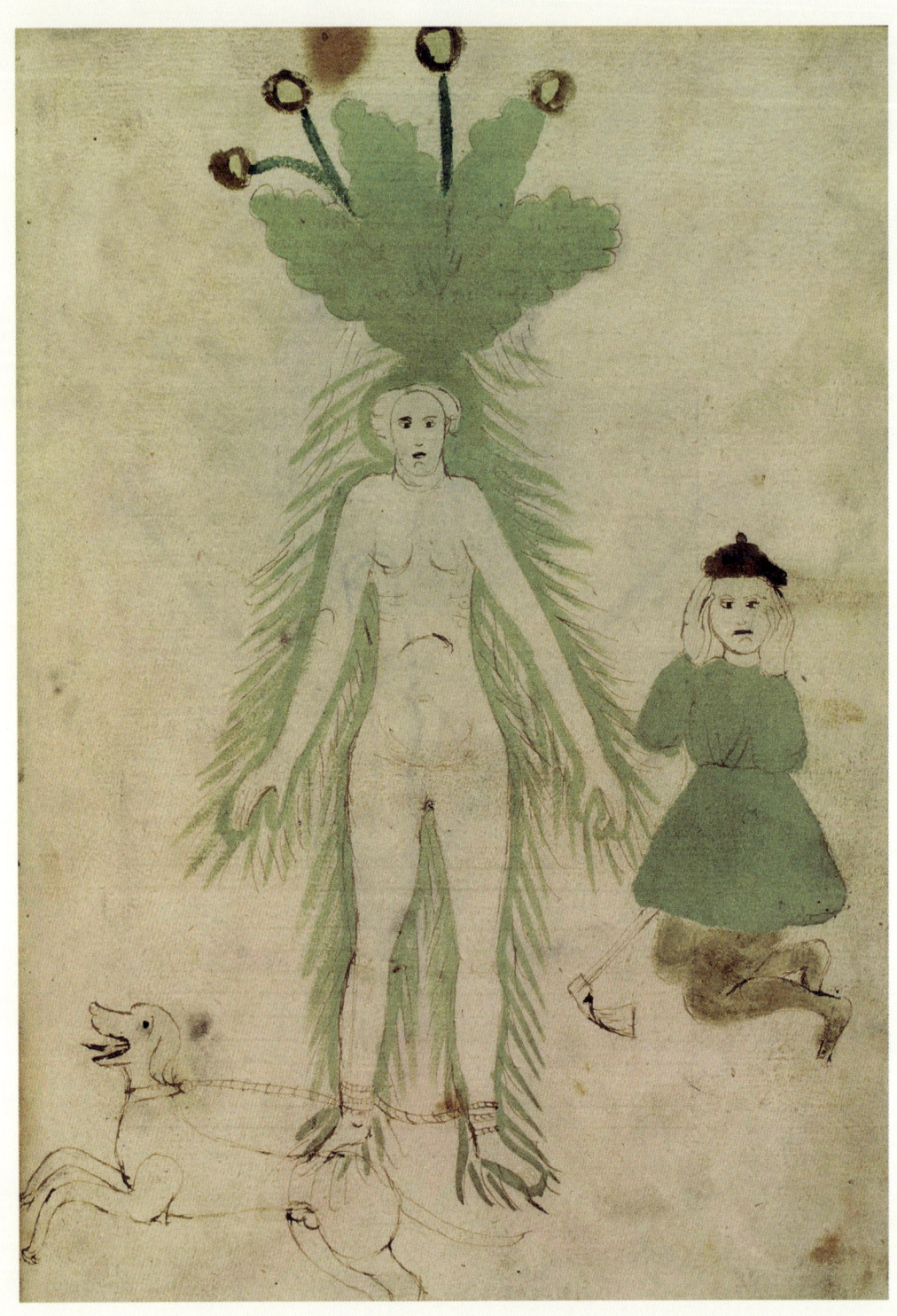

It was thought that mandrakes were inhabited by dangerous creatures called *mandragoras* (dragons in human form), which were commanded by sorcerers.

that scopolamine, one of the alkaloids present in mandrake, can reduce depressive symptoms, so this may be the link to the plant's association with suicide. It is said to have grown alongside Prometheus as he lay chained and being devoured by an eagle, punishment meted out by Zeus for having given humans fire. Mandrake also played a vital role in the lives of some of ancient Rome's most famous figures. In the 2nd century BCE Hannibal of Carthage left out wine laced with mandrake for his enemies, and he returned to regain the city as his enemies lay stupefied. Similar stories are told of Julius Caesar escaping Sicilian pirates in the 1st century BCE.

The origin of the mandrake's name *mandragora* is obscure. It may derive from the Sanskrit *mandra* (sleep) and *agora* (substance). But these same words in Greek mean 'stable' and 'gathering place', which are places where the plant was commonly found. Whatever the case, the plant, in limited quantities, has proven to be a helpful sedative and painkiller. Dioscorides explained how an anaesthetic wine could be made from the mandrake for those suffering from illness or insomnia, or undergoing surgery. It was applied using a sponge, hence its name *spongia somnifera* (soporific sponge), and was used from ancient Greek times well into the Middle Ages. A tincture was made from a mixture of mandrake, henbane (p. 54), hemlock (p. 62), opium (p. 82) and other herbs, absorbed by a sponge and left to dry. During operations the sponge would be rehydrated and held to the nostrils of patients. This acted as an anaesthetic and was said to give relief from pain. It has even been suggested that Jesus Christ was offered *spongia somnifera* as a sedative during the Crucifixion.

Dioscorides' work was translated into Arabic during the expansion of the Islamic empire. In his *Treatise on Poisons and Their Antidotes*, the 12th-century Jewish scholar Maimonides wrote of its use in sleeping pills, and elsewhere he mentioned it as a treatment for asthma. Extracts have also been broken down to make mandrake oil, which is used for treating pain, inducing sleep and improving libido. In the Hebrew Bible *dûdā'im* (love plant) is referenced in Genesis 30: 14–16. This is said to be the oldest reliable mention of mandrake, although a mandrake-like plant was also referred to on ancient Mesopotamian and Egyptian artefacts. The fruit of the plant, commonly called 'love apple', was used as an aphrodisiac and said to enhance fertility. It is reported that the fruit is milder than the root and was cooked with other spices before consumption.

The lore surrounding the mandrake eventually caused its decline in medicine, as did better knowledge of its poisonous properties. But it is still employed in some parts of the world: in Turkey to treat haemorrhoids, gynaecological problems and inflammation; in Italy for boils; and in Jordan to relieve flatulence and rheumatism. In Spain the fresh leaves are used to treat chilblains, and in Israel and Morocco the ancient tradition of using mandrake as an anaesthetic has survived to the present day.

Iron heart

VERVAIN *VERBENA OFFICINALIS* (VERBENACEAE)
COMMON USES ANTIFUNGAL • ANTI-INFLAMMATORY

Verbena is native to Europe but is now also found in temperate areas of Europe, North and South America, Africa, Asia and Australia. It can be confused with the closely related lemon verbena (p. 136), but, although both belong to the same family, they have different medicinal properties.

The common name 'vervain' derives from the Celtic words *fer* (iron) and *faen* (stone). Many of the plant's various European common names include the chemical element iron: in German, *echtes Eisenkraut* (true iron herb), in Dutch *IJzerhard* (iron hard), while the Finnish *rohtorautayrtti*, Danish *læge-jernurt* and Slovak *zelezník lekársky* all translate to 'medical ironherb'. Despite these names, vervain does not contain iron, nor is it an indicator of iron-rich soil. Rather, the herb was believed to strengthen metal if it was added during smelting.

Officinalis comes from the Latin and indicates the plant's use in medicine; indeed, vervain has been used in folk medicine for thousands of years. It was a sacred plant for the ancient Egyptians, who believed that the first vervain plant grew from the tears of Isis as she mourned the passing of her brother and husband, Osiris. In ancient Greece, vervain was called *hierobotane* (holy plant), and in ancient Rome it was *herba sacra* (sacred herb). In Christian countries, the 'herb of the Cross' was believed to steady the bleeding of Jesus's wounds during the Crucifixion.

In the Middle Ages, the plant was used as protection from evil magic, giving rise to the saying: 'Vervain and dill, hinder witches from their will.' It was also used by the Druids, as the historian and philosopher William Winwood Reade wrote in 1861 in a study of these mysterious Celtic figures: 'The vervain was to be gathered at the rise of the dog-star, neither sun nor moon shining at the time; it was to be dug up with an iron instrument and to be waved aloft in the air, the left hand only being used. The leaves, stalks and flowers were dried separately in the shade and were used for the bites of serpents, infused in wine.' Similarly, Hildegard of Bingen described vervain as 'cooling' in her treatise *Physica* (1150–58) owing to its use in treating inflammation, particularly around the mouth.

Vervain's capacity for healing is recognized today. It is a key component in herbal medication used to treat sinus infections. Dried vervain leaves can be made into a tea to treat a range of ailments, including digestive problems, insomnia, pain, anxiety and depression. Vervain is currently used in folk medicine to treat kidney and respiratory troubles, but further scientific study is required to confirm its efficacy; recent studies have produced promising results regarding its antibacterial, antifungal and antioxidant properties. Although cases are few, contact dermatitis and anaphylaxis have been recorded from contact with species of *Verbena*.

During the Middle Ages healing herbs were called 'simples'. Vervain was used so commonly that it gained the name 'simpler's joy'.

In the 17th and 18th centuries alchemists believed that the dew on the leaves of this plant could turn metals into gold, so the term for those early chemists inspired its name, *Alchemilla*.

A lady's herb

LADY'S MANTLE *ALCHEMILLA VULGARIS* (ROSACEAE)
COMMON USES GYNAECOLOGY • CANCER

The frothy flower spike of this small plant contains clusters of little yellow-green flowers and is held above round, green, hydrophobic (water-repelling) leaves with serrated edges, on which raindrops are beautifully visible, caught like droplets on an umbrella. Anyone who has grown *Alchemilla* will be familiar with its self-seeding, often into gravel, grass or cracks in paving. The species epithet, *vulgaris*, translates as 'common' and indeed this adaptable plant is widespread in Europe.

The name *Alchemilla* comes from the medieval Latin *alchymia* (alchemy), an early form of chemistry. Those who practised alchemy believed that the dew on the leaves of lady's mantle was the purest form of water and could turn base metals into gold. While that idea was subsequently disproved, the plant continued to be greatly valued. The word alchemy, meanwhile, survives in the present to signify the process of turning something common and of little value into something of great worth.

The renowned herbalist Nicholas Culpeper wrote in the 17th century that *Alchemilla* was 'highly prized and praised' for its ability to heal wounds and cuts. A herbal monograph published by the European Scientific Cooperative on Phytology in 2013 bore out this claim; when treated with *Alchemilla* under test conditions, minor mouth ulcers showed a marked improvement. A study carried out in 2019 produced similar results.

Different species of *Alchemilla* have been used in medicine, and the common name 'lady's mantle' is generally used for all species in the genus, making it difficult to ascribe specific medicinal properties to particular species. Taxonomically, many species in this genus are now considered subspecies, and the inclusion of microspecies has led to further confusion.

Alchemilla in its various forms was traditionally used as a herbal medicine in southern Europe and the Balkans. It has long been associated with women, as its common name indicates, and used to treat such gynaecological disorders as heavy and painful periods, yeast infections and perimenopausal symptoms, including hot flushes and night sweats. Recent studies have provided strong evidence of anti-cancer properties, particularly in breast and ovarian cancer, a benefit that has been attributed to a concentration of the enzymes amylase and tyrosinase. The American herbalist Gail Edwards recommends lady's mantle in a tea or poultice.

It is now also possible to interrogate the assertion by Elizabeth Blackwell, author of *A Curious Herbal* (1737), that *Alchemilla* leaves could be applied to 'flagging breasts' to improve their firmness. Today, lady's mantle is often used in pharmaceuticals, including moisturisers, sun creams and anti-ageing creams. The salicylic acid in the plant conditions and tightens skin, and has also been used to treat acne.

Herbal dental care

HENBANE *HYOSCYAMUS NIGER* (SOLANACEAE)
COMMON USES POISON • NARCOTIC

Henbane grows in dry, rocky areas in many Mediterranean countries. This tall plant (up to 80cm/32in) with attractive, patterned flowers and arrow-shaped leaves is most commonly an annual or biennial. The biennial (two-year) plant will produce only a leafy rosette in the first year; it then flowers during the second year, producing the next generation of seeds before dying.

The name 'henbane' (hen killer) reflects the fact that chickens were observed eating the plant's seeds, then becoming paralysed by the toxins they contain, and subsequently dying. Pigs, however, appear to be immune to the toxins; this is perhaps why the genus name comes from the Greek *hyos* (pig) and *cymos* (bean), producing another common name, 'hog bean'. The plant was historically used in hunting hares in England and France. A mixture of blood and henbane leaves was spread on the pelt of a dead hare, enticing other hares to come to its aid; after licking it, they would be poisoned and easily caught. In the 15th century henbane seeds were used throughout Europe to catch birds by being mixed with grain, disabling the birds and allowing them to be caught with the bare hands.

In Greek mythology, the gates of Hades were guarded by dead souls wearing collars of henbane. The ancient Greeks often burned it as incense, and the Oracle in Delphi is said to have burned the seeds, inhaling the fumes so that she would see her visions more clearly. As early as the

1st century CE, Dioscorides advocated for the use of henbane as an anaesthetic oil. But Pliny argued that henbane reduced brain function. Henbane – along with mandrake (p. 46), opium (p. 82) and other herbs – were key components of *spongia somnifera* (soporific sponge; see p. 49). The same herbs were used to make a sleep remedy called 'pomander' or 'dwale', which would knock out a patient before surgery. It was critical to get the volumes of plants correct, however, or the mixture could be deadly.

Over time, people have created new ways to contend with henbane's toxicity. In medieval Britain, as in ancient Delphi, witches reportedly burned the plant and inhaled the fumes to welcome spirits and demons when performing conjuring magic. Mixed with fat, it was made into a paste to be spread on the skin. This reduced the risk of poisoning – compared to taking it orally – and increased the rate of absorption.

The use of henbane *for* – not despite – its debilitating effects is fascinating. Ingesting henbane produces a dry mouth, an abnormal heartbeat and a range of neurological problems, such as hallucinations and paralysis. This can lead to coma or death. In Germany henbane (known there as *Bilsenkraut*) was used in beer to enhance the feeling of intoxication, until it was outlawed in 1953.

Seeds discovered in 2002 during an archaeological dig in a Viking grave in Fyrkat, Denmark, are thought to be

Traces of henbane have been found at Neolithic burial sites in Scotland,
suggesting that this herb was important in funeral rituals.

According to Greek mythology, the dead would walk the underworld wearing
crowns of henbane, a herb that was supposed to eradicate all memories.

of henbane. The plant is not native to Scandinavia, so it is clear that the Vikings found it somewhere on their travels. It has consequently been suggested that henbane was used by Viking berserkers (warriors) to intoxicate themselves into a rage fit for battle. There is also evidence that the seeds were used in pain relief, a property the Romans had discovered; they used henbane seeds to treat toothache, hence its name *herba dentaria*, tooth-herb.

Henbane was also reported to cure 'tooth worms', said to be parasites that would gnaw the teeth and cause them to fall out. Widespread belief in tooth worms began as early as the 4th century and continued into the modern age in Europe, and henbane seeds were sold to people suffering from or in fear of the worms. Yet the numbing effect of the seeds gave only temporary relief. Tooth worms were regarded as nonsense by some herbalists, among them John Gerard, who believed the whole idea was just a money-making scheme by charlatans. Indeed, science has now disproved the notion as superstition and make-believe.

In the fictitious Denmark of Shakespeare's tragedy *Hamlet* (I:v), it is thought to be the deadly 'hebona' that poisoned and killed Hamlet's father. Similarly, Catherine 'La Voisin' Deshayes, the infamous 17th-century poisoner – believed to have been involved in between 1,000 and 2,500 murders – is rumoured to have used henbane to perpetrate her crimes.

In London in 1910 one Dr Hawley Harvey Crippen purchased a large amount of henbane leaves, then used to treat Parkinson's disease and to ease childbirth. His wife – who had found out that he had a mistress and threatened to leave him – subsequently disappeared, to be found later by police, buried under cement in the cellar of the couple's house. Despite fleeing the country, Crippen was caught and sentenced to death by hanging for the murder of his wife.

Owing to the deadly range of alkaloids it contains, henbane is hardly used today. There is evidence that the plant was being used in the United States in the 1950s as a pain reliever, for swelling, and as an antidote to mercury poisoning and rodent poison. In Turkey the plant is still employed for toothache, earache and eye irritation; the fumes produced by heating the seeds in a pan are inhaled, and water is added to increase the steam and smoke.

Mediterranean cucumber

BORAGE *BORAGO OFFICINALIS* (BORAGINACEAE)
COMMON USES MEMORY • CULINARY

Borago officinalis originated in the Mediterranean, northern Africa and Asia Minor, but is now naturalized in many other countries, including Britain, most likely owing to the expansion of the Roman empire. This plant grows easily from seed and can be annual or biennial. The flowering stems curve like a scorpion tail and hold purple-blue star-shaped flowers, from which another common name, 'star flower', derives.

The name 'borage' derives from the Latin *cor* (heart) and *ago* (to aid), and refers to the plant's traditional use as heart medicine. The Celtic people named it *barrach* (brave man), and added it to wine to give warriors courage in battle. The German encyclopaedia *Hortus Sanitatis* (1491) suggests that borage was consumed with wine to fend off melancholy. This may indicate why Pliny referred to the plant as *euphrosinum* (from the Greek *euphron*, cheerful) and why John Gerard later described borage being used 'Because it maketh a man merry and joyful'.

Borage has also been cultivated for culinary use. In Liguria, Italy, it is added to pasta fillings, and in Germany, particularly Frankfurt, it is combined with other herbs to make a soup called *grüne Soße* (literally 'green sauce'). In Persian cuisine, a honey-sweet tea is made from the dried flowers. Their enticing taste and delicate beauty have made borage flowers increasingly popular for decorating desserts and for serving with gin.

Medicinally, *B. officinalis* has traditionally been used to treat several conditions, among them scurvy, jaundice and inflammation. It is still used for respiratory problems, urinary tract infections, arthritis and skin diseases, although it is best known as a dietary supplement. A study carried out in 2001 found that borage can improve memory, particularly in people over ninety years old; other studies have reported that it may assist in treating obsessive compulsive disorder and seizures (although yet other studies have found that too much borage can actually induce seizures).

Borage's commercial use is focused on the nutritional supplement borage-seed oil. Rich in GLA (gamma-linolenic acid), an omega-6 fatty acid, it has been used historically to treat atopic eczema, and is now being studied as a potential treatment for such inflammatory disorders as arthritis and asthma. However, some members of the borage family, including borage itself, contain the alkaloid pyrrolizidine, which has been linked to problems with liver function.

Borage flowers were candied and eaten as sweets
during the first Elizabethan era in England.

Hollyhock is an attractive, easily grown garden plant, and a large
range of cultivars is commercially available.

The holy hock

HOLLYHOCK *ALCEA ROSEA* (MALVACEAE)
COMMON USES COAGULANT • SWELLING

Alcea (sometimes spelled *Althaea*) is a genus that is present in Europe but has uncertain origins. The theory that it was brought to Europe from southwestern China in the early 15th century by the Ming dynasty explorer Zheng He is unlikely, since the species is not known in the wild in China. The 20th-century plant geneticist Daniel Zohary claimed that *Alcea* may have originated in the Aegean Islands and the neighbouring Balkans, an assertion that recognizes the largely eastern European and eastern Asian distribution of the genus.

Alcea is in the family that also contains such familiar plants as hibiscus (*Hibiscus* spp.), mallow (*Malva* spp.) and cocoa (*Theobroma cacao*). The genus name comes from the Greek word *alkaia* (mallow). The brightly coloured flowers and height of this tall plant make it popular in gardens. It is typically an annual but can be perennial, growing best in dry locations, such as in the cracks of pavements and beneath walls. The species name, *rosea*, comes from the flower's colour; hollyhock petals provide a red pigment that can be used to colour preserves. In truth, the flower colour is variable, and many selections in different colours have been made. The common name 'hollyhock' can be traced to the 16th-century herbalist William Turner, who referred to 'holyoke' – from 'holy' and the old English word for mallow, 'hoc'. The plant's numerous healing properties were the basis of its holiness.

Many of the mallows have symbolic meanings that relate to romance, fertility and the circle of life, perhaps because of the plant's vigorous growth and prolific flowering even in very dry situations. It is thought that in ancient Egypt mallow flowers were buried with the dead to help them be guided to the afterlife.

Hollyhock is a very old medicinal plant. It was used by the Neanderthals about 60,000 years ago for healing and rituals, and later became an important ingredient in Chinese traditional medicine. In the Xinjiang region of China it is used to treat inflammation, to reduce fever, and as a diuretic. The Uyghur people use the flowers to stop bleeding, reduce swelling and detoxify the body, and the seeds are used to treat inflammation of the kidneys and uterus. The plant's medicinal properties have a wider geographical reach, though: in Iran, the flower is used to treat swollen gums and as a laxative; and in Saudi Arabia, the leaves, branches and fruit form part of a worming agent for cows, camels and sheep.

The petals can also be infused into a herbal tea. All parts of the plant – petals, buds, new leaves and inner part of the young stems – can be eaten raw in salads. Beautiful and also useful, this surprisingly multifunctional plant is one of great cultural significance.

Mark of death

HEMLOCK *CONIUM MACULATUM* (APIACEAE)
COMMON USES POISON · WITCHCRAFT

Hemlock is one of the most toxic plants in the world. Eating it leads to gradual paralysis: breathing slows and the organs shut down, but in fact death comes from asphyxiation, courtesy of the chemical compound coniine. The genus name *Conium* comes from the Greek *konas* (to whirl), a strangely incongruous reference to its effect on the body. Hemlock was used to execute criminals in 5th-century BCE Athens, and it was part of the concoction that fulfilled the philosopher Socrates' death sentence in 399 BCE.

The plant originated in Europe and was introduced to Norway alongside grain in the 1920s and 30s by the Soviet Union. While it is most common in damp soils with high levels of nitrogen, hemlock can grow in a range of habitats and is now spread across the temperate regions of the world. A single plant can produce up to a staggering 40,000 seeds, making it challenging to eradicate once established. Hemlock is often responsible for killing livestock, particularly cows, and human poisoning also still occurs, mostly affecting those who misidentify the plant as something edible, such as parsley (p. 21).

The word 'hemlock' comes from the Anglo-Saxon *hemlic* or *hymelic*. The spelling varied until Shakespeare's play *Henry V* (V:ii), when it became immortalized as hemlock. The species name, *maculatum*, refers to the characteristic spotted stalks of the plant (an important distinction between it and parsley, for example). These markings explain another common name for hemlock,

'mark of Cain', a reference to the biblical story in Genesis 4:15 in which Cain (the son of Adam and Eve) kills his brother Abel out of anger. He lies when confronted by God and is marked as an example to others of the consequences of taking the life of another. The indicator of deadliness is certainly an apt one in the case of hemlock.

Predictably, given its use in poisons, hemlock has also historically been referred to as the 'devil's flower'. It was linked in the Middle Ages to witches and their creations of spells, hexes and poisons – 'the devil's work' – alongside mandrake, henbane and deadly nightshade (pp. 46, 54, 81).

Despite hemlock's well-known toxicity, it has been used regularly for its hallucinogenic properties. The high concentration of alkaloids can produce disturbing hallucinations and have lasting effects on the central nervous system. Greek and Arabian physicians used hemlock to treat tumours and swelling. The anaesthetic drink known as 'dwale' – thought to have originated in ancient Rome, and made from hemlock, opium (p. 82), henbane, mulberry juice, lettuce, mandrake and ivy – was used widely across Britain between the 12th and 15th centuries. Towards the end of that period, the roasted roots were used to relieve the pain associated with gout, and the unripe seeds were dried and used as a sedative. Strangely enough, hemlock was also administered by the ancient Greeks as a last-chance antidote for severe poisoning.

Despite its poisonous reputation, hemlock is a good food plant for moth larvae,
particularly those of the poison hemlock moth.

The leaves and flowers of agrimony can be used to make a yellow dye, the colour strength of which depends on the time of harvesting.

The emperor's herb

AGRIMONY *AGRIMONIA EUPATORIA* (ROSACEAE)
COMMON USE DIGESTIVE COMPLAINTS

This native of Europe and southwestern Asia grows in marshy areas or on the banks of streams, standing roughly 1m (3ft) tall, with fluffy, feather-like tufts of green leaves. The long flower spike holds a stream of yellow flowers that produce lots of pollen, making them attractive to honeybees, hoverflies and butterflies. Once fertilized, agrimony produces seeds with small hooks that attach to the fur of passing animals, an adaptation that ensures their wide dispersal.

The name *Agrimonia* comes from the Greek *agros* (field) and *monas* (lonely), referring to the fact that it is not keen on too much competition from other plants. The species name *eupatoria* comes from the Anatolian ruler Mithridates VI Eupator, ruler of the kingdom of Pontus in northern Anatolia in the 1st century BCE and one of the Roman empire's greatest foes. He in turn is the source of the word 'mithridatism': the practice of repeatedly taking non-lethal amounts of poison in order to develop immunity. Mithridates' father was poisoned at his own banquet, and Mithridates feared meeting the same fate, so he took small amounts of agrimony, among other poisons, in order to protect himself from any substances that others might attempt to administer.

Agrimony has a long history of use by people. Pliny thought very highly of it, calling it 'a herb of princely authorite'. This is probably a reference to its association with Mithridates. The Anglo-Saxons called it *garelive* and considered it sacred; they used it to treat wounds, warts and snake bites.

In medieval France, agrimony was the main ingredient in arquebusade water, a coagulant used to slow or stop the bleeding of wounds on the battlefield. Consequently, soldiers called it the 'gunshot herb'. In Scotland, meanwhile, it was recorded as a 'witch's cure' for unexplained illness. This may have roots in shamanic herbalism, which is based on the belief that a person can influence good and evil spirits in the world. The tea is drunk to help banish unwanted spirits or to exorcise a spirit that has taken hold. Burning the plant as incense was supposed to ward off evil energy, break spells and remove hexes. More prosaically, Nicholas Culpeper recommended agrimony to treat gout.

Agrimony's use as a herbal remedy continues today. It is used to reduce bleeding, improve skin health, promote hair and nail strength, and to treat liver, kidney and gallbladder problems. Infused as a tea, it is known to reduce inflammation, ease painful coughs and treat diarrhoea. In some European countries, such as Italy, agrimony is used in a tonic to improve digestion. Looking to the future, it has been suggested that this herb has antiviral properties that could be employed against the hepatitis B virus. Additionally, research is being conducted into its ability to aid diabetic patients, since some of its active compounds may improve glucose regulation and the uptake of insulin.

Angelic angels

ANGELICA *ANGELICA ARCHANGELICA* (APIACEAE)
COMMON USE CULINARY

This plant is heavenly in every way. The 'herb of the angels' has a beautifully sweet scent, and is highly regarded as an ornamental thanks to its graceful habit and size. *Angelica* is Latin for 'angelic', and *archangelica* comes from the Greek *arkhangelos*, the highest order of angel. Indeed, the English herbalist John Parkinson's *Paradisi in sole paradisus terrestris* (a punning title meaning Park-in-sun's Earthly Paradise; 1629) puts it as one of the most highly regarded herbs.

This biennial plant – which produces only a rosette during its first year, conserving its energy to produce the huge 2.5m (8ft) flowering stem of its second year – grows in northern countries: Scandinavia, Russia, Greenland and Iceland. The Vikings are understood to have consumed it raw while on marches, and used it to complement other foods, such as fish. They are also credited with extending its native range and introducing it to Scotland, southern England and Ireland. The plant went unmentioned in the medicinal books of Greek and Roman physicians, but was grown by medieval monks. It is said that the archangel St Michael visited a monk in a dream and told him of the wonders of this plant. He said it could be used to treat victims of bubonic plague, the 'Black Death', which killed between 75 and 200 million people in 1347–51. Angelica – the 'root of the Holy Ghost' – could apparently also repel witchcraft, enchantment and evil spirits. The root, however, is poisonous when fresh, and must be dried if it is to be made safe for use and consumption.

The cultivation of angelica increased in northern countries. As the crop developed, selections were made, including the 'Vossakvann' selection in Voss, western Norway. Its filled-in stems resulted in a higher yield and a better taste.

The Vikings and the creators of the Norwegian selection were far from the only people to recognize angelica's culinary potential, however. Famously, the 10th-century Norwegian king Olaf Tryggvason, the first king to allow Christianity in Norway, is depicted in a drawing by the late 19th-century artist Erik Werenskiold trying to appease his wife by offering her angelica. According to the 18th-century Swedish botanist Carolus Linnaeus, the Sámi people of northern Scandinavia ate the sweet stems 'like an apple'. In Greenland, the stem is eaten with powdered sugar, especially by children, and in the Faroe Islands it is served with cream or thick, sour milk. Candied angelica is the most famous use of the herb in Nordic cuisine. In the Poitou-Charentes region of western France, it is used in the dessert liqueur *fleur d'angelique*.

Angelica is useful in medicine, containing phytosterols, saponins and gelatinoids that can interact with hormones to treat hot flushes, menstrual cramps and mood swings. It also has a range of antioxidant-rich minerals and vitamins, which are useful for relieving nerve and muscle tension, and for improving sleep.

Angelica is often used as an aromatic in gin, since it pairs well with juniper berries.

Chewing common rue leaves – something the artists Michelangelo and Leonardo
da Vinci did regularly – was believed to improve the eyesight.

Rue the day

COMMON RUE *RUTA GRAVEOLENS* (RUTACEAE)
COMMON USES SPICE • CAT REPELLENT

Ruta graveolens is an evergreen shrub from southern Europe that grows best in well-drained soil. The aromatic leaves are an attractive blue-green and form a globe as the plant establishes. However, it is important to note that all *Ruta* species can cause contact phytophotodermatitis (a condition that can increase the skin's sensitivity to light, resulting in sun damage), so should be handled with gloves. The origin of the common name 'rue' is unclear, although it is often associated with the English word for regret. The phrase 'rue the day' may be related to the difficulties of handling the plant, representing the misgivings of someone who did not adequately protect themselves from risk.

Common rue was well known to the Greeks and Romans. Dioscorides wrote of a range of medicinal uses, including the treatment of urinary tract infections and headaches. Aristotle reported that the plant could ease tension, and Pliny remarked that painters used it to improve their eyesight (probably because of the flavonoids the plant contains, notably rutin). In the early 3rd century CE the Greek writer Aelianus contended that a weasel bitten by a snake would retreat and eat rue to numb the effects of the venom.

In the Middle Ages, the plant gained a new common name: the 'herb of grace'. During outbreaks of plague, it was mixed with holy water and spread around churches, cathedrals and courtrooms to prevent the disease from spreading. Rue was also reported to be used in exorcisms and as protection against spells. This is perhaps how it obtained its Latin name; *Ruta* comes from the Greek word *reuo*, to set free. Rue was grown as a tribute to the Virgin Mary and was strongly associated with purity and maidenhood. Catholic missionaries collected rue and took it to Lithuania, where it became the country's national flower – unusually, since national flowers are typically native species.

The species name, *graveolens*, references the intense smell of the plant's leaves, a property that makes it one of the best cat repellents. As with other plants that contain strong essential oils, it can be especially overpowering to cats' sensitive noses.

Rue can also be made into ruda tea, which is unsafe during pregnancy because it is said to stimulate menstrual flow. It is also, like green tea, high in the antioxidants known as catechins, which prevent the body from absorbing the essential chemical folic acid. It can thus result in defects in the neural tube of a foetus, leading to severe problems with brain development.

The bitter taste of the leaves means that rue is largely unused in European kitchens. However, it is a key ingredient in Ethiopian and Eritrean cuisine, and features in the spice blend known as *berbere*, which translates as 'pepper' or 'hot'. Berbere is a key component in the famous *doro wat*, an aromatic chicken stew served with *injera* (fermented bread made with teff flour).

Deadly nightshade; see p. 81.

NARCOTICS AND POISONS

Narcotics and poisons

Herbs are often thought of in a positive way, but they also have a dark side. Alongside the many healing herbs, there are others that can upset the stomach, cause hallucinations and even kill. These dangers are the result of adaptations that prevent animals from eating the plants. But even some of these dangerous species can be beneficial if administered in low doses or used specifically. Foxglove (p. 94), for example, is useful in medicine and even saves lives. The line between heal and harm is thin, however, and overdose can prove deadly.

Medicine has always struggled to strike a balance between good and bad effects. Roman herbalists were as much killers as they were healers, and the quest for power and riches drove some people to conspire to murder their opponents and even, at times, their allies. Of all assassination methods, the herbal approach was among the most popular. Poisoning was rife in the Roman empire; many affluent people lived in fear of their drinks and food being poisoned, and rulers demanded the identification of all possible poisons and their antidotes. If they fell victim to poisoning, death was not instantaneous, so the perpetrator could make their escape and concoct an alibi. In those times, before autopsy, many murders went not only unsolved, but also unsuspected.

Poisons are not commonly used today, but the use of narcotics is constantly debated. Since the mid-20th century, legislation on drugs has become more punitive, punishing both dealers and users. Wars on drugs have erupted across the world as narcotics are grown and shipped globally via underground networks – the present-day equivalent of the historically important Silk Road trade route. Today the production of narcotics is one of the main sources of income for organized crime, and has direct links to modern slavery, violence and drug addiction.

To suggest that all narcotics are bad would be an oversimplification. Cannabis (p. 90) is the best-known plant to be caught in this ethical debate. It has a long history of use in clothing and other goods, and for food, but has been vilified for years. Since the turn of the millennium, however, it has been legalized for recreational and medicinal use in many parts of the world. Other herbs, among them jimson weed, coca and tobacco (pp. 77, 78, 86), had been used recreationally by Indigenous people for centuries without becoming a societal problem.

Therefore, it is the way humans use it that makes a herb 'bad'. Once synthesized into strongly addictive drugs – opium (p. 82), tobacco and cocaine, for example – they become very different from the natural plant. These artificial drugs are then exported across the world, making them available to millions who would not otherwise have accessed them. In the story of narcotics, greed has always been valued over health. Studies have been silenced, and when questions have been raised, they have been ignored.

The story of poisons is similar. There is a difference between being venomous and being poisonous. Venom is inflicted on a victim by a predator, while poison is purely a defensive mechanism employed by prey. Poisonous plants have evolved to discourage humans from ingesting them; they are not predators but prey. It is humans who are responsible for weaponizing poisons and using them for unjust and unlawful reasons.

In this chapter, I have made a diverse selection of narcotics, from tobacco, which is commercially available across the globe, to cocaine, which is a strongly banned substance. I have included opium for its historical prominence as a drug. Poisons that have been used in medicine, such as foxglove, indicate the precariousness of the balance between herbal remedy and poison. Other herbs, such as deadly nightshade and aconite (pp. 81, 89), are included for their historically murderous use. By detailing both the historical importance and recent appearances of these herbs, this chapter aims to display the complex relationship humans have with these powerful plants.

Artistic liquid courage

WORMWOOD *ARTEMISIA ABSINTHIUM* **(ASTERACEAE)**
COMMON USES DRINK · HALLUCINOGEN

The soft leaves of this small, upright shrub are bluish-grey, making the plant look dainty, but in fact wormwood is strong and can grow in a range of climates. It produces dandelion-like seeds that are carried by wind, a method of dispersal that has allowed it to spread from its native ranges in Eurasia and northern Africa.

The first uses of this herb are recorded in the Ebers Papyrus of about 1550 BCE. The Greek philosopher Theophrastus in the 3rd century BCE referred to its bitterness and declared it unfit for consumption, but in the 1st century CE Dioscorides mentioned its use for digestive problems. At about the same time, Pliny described how, when the plant's ashes were mixed with rose petals, 'it blackens the hair' – potentially the world's first mention of hair dye. Hildegard of Bingen's mid-12th-century treatise *Physica* refers to wormwood as 'the most important master against all exhaustions'. However, by the 18th century, use of the plant had effectively stopped. It was then that the essential oils were extracted, first in Switzerland, and used to make a powerful and controversial alcoholic drink: absinthe.

Wormwood, anise (*Pimpinella aromatica*) and fennel (p. 18) are the main ingredients of absinthe. During distillation, the plant is broken down and the chlorophyll extracted. This gives the drink its signature luminous green colour, prompting the informal name 'the green fairy'. Absinthe was extremely popular in the 19th century, especially in France, and it was later sent to French soldiers fighting in Algeria in the 1950s and 60s to protect them from malaria and dysentery. It was a particular favourite of writers and artists. Oscar Wilde, Alfred Jarry and Pablo Picasso are all known to have enjoyed it, and the American writer Ernest Hemingway described it beautifully as 'opaque, bitter, tongue-numbing, brain-warming, stomach-warming, idea-changing liquid alchemy'. It is believed that Vincent van Gogh made his famous painting *The Starry Night* (1889) under the influence of a delirium induced by absinthe.

The drink soon became associated with hallucinations and violent behaviour, however, and several countries banned it in the early 20th century, among them its country of origin, Switzerland. It was later discovered that the dangerous effects were not necessarily caused by the drink itself, and since the 1990s absinthe has become available much more widely once more.

Wormwood is still used for a range of purposes. Studies of its essential oils have revealed antibacterial, antifungal and antiprotozoal activity (protozoa being the parasitic organisms responsible for such diseases as malaria). Research increased dramatically after annual mugwort (*A. annua*) was found to contain artemisinin, the discovery of which as a treatment for malaria earned the Chinese scientist Tu Youyou the Nobel Prize in Physiology or Medicine in 2015. Perhaps this emphasizes how much we could still learn about, and from, wormwood.

Wormwood prefers dry, nitrogen-rich soil. Its delicate appearance
belies its resilience to tough growing conditions.

In Australia in 2022, *Datura* contaminated some spinach crops, resulting in 200 people becoming ill.

Jamestown poison

JIMSON WEED *DATURA STRAMONIUM* (SOLANACEAE)
COMMON USES HALLUCINOGEN • PAINKILLER

Jimson weed – also called 'thorn apple' and 'devil's trumpet' – is an annual plant that has been cultivated so widely across the world that its exact origin is difficult to pinpoint. Some believe it came from the Americas, since the highest concentration of plants is found in Mexico and Central America. Others believe it is native to the region surrounding the Caspian Sea, between Europe and Asia, but this may simply be because the first jimson weed seeds to be grown in England came from Constantinople (present-day Istanbul).

The common name refers to the location of several poisonings in 1676: the American settlement of Jamestown, near what is now Williamsburg, Virginia. It is said that British soldiers during Bacon's Rebellion – an uprising led by one Nathaniel Bacon against the governor of the colony – mistakenly ate jimson weed seeds, and suffered days of hallucinations as a result. (The plant's hallucinogenic properties are well known, and have earned it the name 'locoweed'.) The story caught the public imagination, and over the years 'Jamestown' became corrupted to 'jimson'.

Eating jimson weed causes symptoms comparable to those brought about by henbane (p. 54), producing a trance that can lead to coma. Death is rare, but survivors can experience severe memory loss. Nancy Hanks Lincoln, Abraham Lincoln's mother, may be one of the most notable victims of jimson weed poisoning. She drank milk containing extracts of jimson weed and/or snakeroot (*Ageratina altissima*) and developed 'milk sickness'. The contamination of milk with jimson weed continues to be a problem in South Africa, and in Afrikaans the plant is called *malpitte* (evil seeds).

Yet jimson weed is unquestionably attractive. Since at least 1500, it has been cultivated – mostly as an ornamental – in Italy and Germany. The plant produces a white, trumpet-shaped flower and later seeds in a spiky capsule. Its appearance caught the eye of the artist Georgia O'Keeffe, whose painting of it from 1936 sold for $44 million dollars in 2014, making her the highest-selling female artist in history.

Jimson weed is an important plant to some Indigenous communities. The Chumash people of California, for example, use its hallucinogenic effects during puberty rituals. The plant has not been used widely in Europe, owing to its relatively late introduction and its similarity to related genera, such as mandrake (p. 46) and henbane.

The genus name *Datura* is said to come from Sanskrit, and the plant is listed in the 6th-century BCE collection of Ayurvedic medicinal texts known as the *Sushruta-Samhita*. In modern Ayurvedic medicine in India, the plant is used to treat ulcers, toothache and swelling. According to Taoist legend in China, the behaviour of the collector of the plant passes to whoever drinks its infusion.

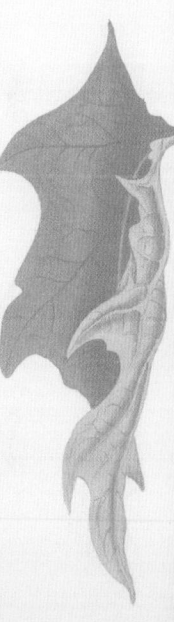

White powder

COCA *ERYTHROXYLUM COCA* **(ERYTHROXYLACEAE)**
COMMON USE DRUG

Erythroxylum coca is native to parts of northern South America, and has naturalized in central Africa and the Indian state of Assam. The genus name comes from the Greek *erythros* (red) and *xylon* (wood), referring to the red timber of this tall shrub. Happiest in acid soils, it produces small white flowers followed by red berries.

Coca is one of its native region's oldest cultivated plants, and there its leaves have been chewed by Indigenous people for more than 5,000 years. It was said to reduce hunger and increase endurance, and was also used in religious ceremonies. The plant was of great importance to the Incas, who believed it to have been created by the god Inti, under instructions from the moon goddess Mama Killa. Only descendants of the gods could use it to ease their hunger – and in order to fulfil the wishes of the deities.

During the Spanish conquest of the Inca empire in the mid-16th century, the conquistadors forced the Indigenous people to work in mines, and gave them coca leaves to save food rations. Production of *E. coca* increased as a result, but it was not until two centuries later that the plant was introduced to Europe, by the botanist Joseph de Jussieu. When cocaine was first made from coca leaves in 1860 by the German chemist Albert Niemann, it showed promise as a beneficial drug, and its popularity in Europe and North America surged. It was, for a period, used as a general anaesthetic, but this stopped when its highly addictive

and hallucinatory effects were recognized.

Nevertheless, a wine made using coca leaves, invented by the pharmacist John Pemberton, was common in the United States as a treatment for various ills. When temperance legislation in 1885 prevented the sale of alcohol, Pemberton had to remove the alcohol and focus on using the coca, resulting in an early version of Coca-Cola. After his death in 1888 the recipe was sold on; coca leaves were used in its production until 1903, and – with the cocaine removed – they are still one of the flavourings in this perennially popular drink.

The use of cocaine as a drug came to be associated more and more with crime, particularly that perpetrated by Black people during the uneasy decades following the abolition of slavery in 1865. Black people were persecuted in the media and in 'scientific' journals, and with the Harrison Narcotics Act in 1914, a total of 46 out of the 48 American states banned cocaine. In the 1980s, however, a cheaper and more quickly produced version of powdered cocaine – crack – came into wide use, and addiction to it devastated communities across America.

In 2016, according to the United Nations World Drug Report, between 800 and 1,000 tonnes of cocaine were produced for about 18.7 million users globally. In a study carried out in 2020, that figure had almost doubled to 1,982 tonnes. Long after the Spanish invaders, *E. coca* is still strongly linked to slave labour, trafficking and exploitation.

The use of coca leaves remains legal in Andean nations, where they are made into tea or chewed. They are particularly useful in lessening the symptoms of altitude sickness.

Deadly nightshade was once employed as a sedative for operations, but
it caused so many deaths that this use was abandoned.

Strings of fate

DEADLY NIGHTSHADE *ATROPA BELLA-DONNA* (SOLANACEAE)
COMMON USES WITCHCRAFT • PAINKILLER

This dangerously toxic species was originally a Mediterranean plant, growing in southern Europe and North Africa along to the Caucasus, but it has been spread beyond its native area by people, and has become a noxious weed in some areas. Its black fruit is eaten by birds and dispersed further. Birds appear to be unaffected by its toxicity, while other animals are not so lucky. The alkaloids present in deadly nightshade are also found in the plants of related genera, such as henbane and jimson weed (pp. 54, 77). *Belladonna* is Italian for 'beautiful woman', an allusion to the plant's historical use, particularly in Italy and France, as a cosmetic to dilate the pupils (mydriasis). This practice was understood to make a person more beautiful, yet *Atropa* is much better known for its poisonous nature.

The genus name comes from that of the Greek goddess Atropos, the daughter of darkness (Erebus) and night (Nyx), and the sister of death (Thanatos). Atropos was one of the Three Fates (with Clotho and Lachesis), who would weave threads to decide the course of a person's life. Atropos made the final cut, marking death.

Ingesting deadly nightshade results in a range of symptoms relating to the nervous system, the airways and the heart. Unsurprisingly, then, the herb has a history steeped in death. Pliny records it as being used on the tips of arrows and spears, and it was reportedly used in the murder of several Roman emperors. In the 1950s plasters containing extract of deadly nightshade (and sometimes also that of aconite; p. 89) to dull pain could be purchased over the counter. It is thought that those with murderous intent would buy large numbers of these plasters in order to extract the deadly alkaloid for a very different purpose. In 1966 the occultist Robert Cochrane committed suicide by ingesting the plant. Accidental poisonings have also been recorded, mostly in children or in individuals who have misidentified the plant – which has attractively shiny, sweet berries – as edible.

Deadly nightshade is well known in herbal medicine and folklore alike. It was valued by witches and regarded as one of the devil's favourites. The berries were said to have hallucinogenic properties that made communicating with other worlds and with spirits easier, so witches were understood to eat small amounts, believing that it would improve their fortune-telling. Contrastingly, peasants reportedly kept deadly nightshade in their homes to discourage evil spirits.

Despite its infamy and extremely poisonous chemical compounds, deadly nightshade is used in some countries today. It is administered for painful joints in Greece, for Parkinson's disease in Bulgaria and for asthma in Serbia, and in Ukraine it is mixed with alcohol to treat rheumatism. It was even used to poison fish in Slovakia, where fresh berries are mixed with bread and butter, then thrown into the water. Similar practices have been recorded in Romania.

Seeds of war

OPIUM POPPY *PAPAVER SOMNIFERUM* **(PAPAVERACEAE)**
COMMON USES PAINKILLER • CROP • DRUG

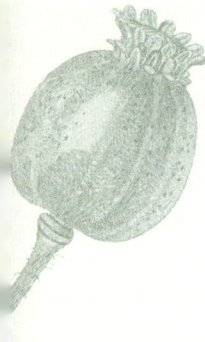

It is likely that the opium poppy originated in the eastern Mediterranean, but its long history of cultivation, export and prolific seeding has made its exact origins difficult to pinpoint. This strong, easy-going annual grows to about 1m (3ft) tall. It is variable in colour, and many garden-worthy cultivars have been selected. It grows in a variety of habitats, from arid ground to lush herbaceous borders.

During a study carried out in 2020 into eleven Neolithic sites in the western Mediterranean, northwestern Europe and the western Alps, remains of opium were dated to between 5622 and 4050 BCE. This suggests the plant was cultivated by Neolithic people. It was probably used during this period for its psychoactive qualities, and as a source of food and flammable oil. It was known to the Minoan civilization (3000–1000 BCE) on Crete and the islands of the Aegean Sea. Relics unearthed in the region in 1959 resembled the shape of the poppy's seed head and were strongly reminiscent of other artefacts found in Greece for the consumption of opium.

Cypriot vases found in Egypt are remarkably similar to these Minoan artefacts, suggesting that the Egyptians were also familiar with opium and its consumption. The opium poppy was mentioned by the name *seter-seref* in the Ebers Papyrus of ancient Egypt as a narcotic and eye medication, and for pain relief. Sleep temples were safe spaces where people could be placed in hypnotic states to be cured of physical and psychological ailments, and to have their dreams analysed, and opium is recorded as being administered for this purpose. According to *The Assyrian Herbal* (1924) by the archaeologist Reginald Campbell Thompson, the plants were known to the Babylonian and Assyrian empires in Mesopotamia. The *pa pa* plant mentioned in a comprehensive Assyrian herbal on stone tablets, excavated by Sir Austen Henry Layard in the mid-19th century and now in the British Museum in London, is thought to date back to 600 BCE, and is likely to be the origin of the genus name *Papaver*.

The first written record of the opium poppy in ancient Greece is found in the work of the poet Hesiod. He wrote in about 800 BCE of the 'poppy town' Mekone, which was presumably known for its production of the plant. The Greek goddess of harvest and agriculture, Demeter, was said to have eaten poppies in order to cope with her grief over her daughter Persephone's kidnapping by Hades, the god of the underworld. This allowed her to fall asleep and wake up refreshed. Opium's sleep-inducing qualities led to its association with Hypnos, the Greek god of sleep and dreams. A drink made of opium and hemlock (p. 62) was created to ensure a quick and painless death, and was often given to older members of the community before they became weak and infirm. This end was described as euphoria, followed

The opium poppy was part of everyday life in China from the early Middle Ages, served at homes and gatherings and in tea shops, in a similar way to present-day tea ceremonies.

Codeine was extracted from opium in 1832 and is still used today as a painkiller.

by muscle relaxation and slowed breathing, culminating in complete respiratory failure.

The opium poppy's journey to Asia is somewhat unclear. It is possible that it was taken to India by the expansion of Alexander the Great's empire in 326 BCE, although no archaeological evidence of its use has been found there. In any case, its use in China and Japan began considerably earlier, between 400 and 700 BCE. It is more likely that it was taken east by Arab traders via the Silk Road.

The English began trading in China in 1715, and in the following century the East India Company held a monopoly on the trade in opium. By the late 1830s the use of opium in China had soared, becoming an epidemic. When the Qing government imposed stricter laws on its importation and sale, the British denied being responsible for selling it to private merchants. After a period of rising tension, the First Opium War (1839–42) resulted in the one-sided Treaty of Nanjing, according to which the British – the victors – were allowed to trade widely and were ceded the territory of Hong Kong.

The word 'opium' comes from the Greek *opos* (juice) and *pion* (fat), a reference to the thick, white sap that oozes from the plant when it is damaged. It is not clear when the drug was first extracted. The highest concentration of opium is in the seed capsules, which are scored to release the sap. This raw opium is boiled and allowed to dry, before being smoked.

Opium contains opiates, which include morphine and codeine, chemicals that have a range of effects, including numbing, soothing and sedation. Opioids are similar substances but synthesized chemically, and they include diamorphine, a very strong painkiller that is used only in extreme cases. Such medicines must be prescribed by a medical professional. Diamorphine was first synthesized by the English chemist Charles Romley Alder Wright in 1874. Felix Hoffmann, an employee at the Bayer pharmaceutical company, carried out the same experiment in 1897. Hoffmann was trying to create a less addictive drug than morphine, but inadvertently he made refined opium, which is up to twice as addictive as morphine. This led to the creation of arguably the most addictive drug in the world: heroin.

At present, opium production covers about 300,000 hectares (740,000 acres) across the world. The largest producers are Afghanistan, India, Myanmar and Turkey. Afghanistan's production has rocketed since the turn of the millennium and now accounts for 90 per cent of global opium supply. The United Nations World Drug Report 2022 showed an increase of 32 per cent in the cultivation of opium since the Taliban took over there in August 2021. The price of opium tripled after the Taliban announced a cultivation ban in Afghanistan in April 2022; the income made by farmers from its sale was valued in that year at US$1.4 billion, up from US$425 million in 2021.

The world's deadliest herb

TOBACCO *NICOTIANA TABACUM* (SOLANACEAE)
COMMON USE CIGARETTES

Tobacco is a quick-growing annual plant that can reach 3m (10ft) in a single season. It is very sticky, owing to glands on the leaves and stems, and produces such an abundance of seeds that it has become a weed in some tropical areas.

Cultivation began some time between 5000 and 3000 BCE in northern South America, and subsequently spread throughout the Americas. The name comes from *tabaco* or *tavaco*, a Spanish corruption of the Indigenous word for the pipe used to smoke the plant; European explorers mistook this for the name of the herb itself. According to herbals, the true Indigenous names for the plant were *petum*, *betum*, *cogioba* and *yietl*. The Portuguese explorer Pedro Alvarez Cabral reported in the 15th century that the plant was used by Indigenous people in Brazil to treat abscesses. It was also used as toothpaste in Venezuela, and in Mexico, the scent of the fresh leaves was said to clear headaches. Indigenous people used tobacco as a body rub to kill lice; they also drank or chewed it to become intoxicated, and used it for 'snuffing': sniffing the crushed, dried leaves to gain a high. This habit later became popular throughout Europe.

It is not certain when tobacco crossed the Atlantic. The Franciscan friar Ramón Pané, who sailed with Christopher Columbus in 1492, is thought to be the source, since he was one of the first Europeans to encounter the plant. Tobacco is listed in European herbals and pharmacopoeias from the 1530s onwards, by which time it had an ever-expanding list of uses and had gained a reputation as a 'holy herb', able to cure all. Ironically, given its later associations, it was said to prevent cancer and lung infections. It was even said to have cured Catherine de' Medici, Queen of France in the mid-16th century, of headaches, and was consequently named *herba regina* (queen of herbs). The genus name, *Nicotiana*, comes from Jean Nicot, the French ambassador to Portugal, who introduced it to the queen.

In the early 19th century scientific attention turned to the plant's chemical make-up. Nicotine was first isolated by the German chemists Wilhelm Heinrich Posselt and Karl Ludwig Reimann in 1828; it was later discovered to be a dangerous alkaloid, fatal at doses greater than 40–60mg. The average cigarette contains 1mg, so smoking many over a short time can cause fatal poisoning. Neonicotinoids are used in insecticides, although this has diminished in recent years owing to the harm they are thought to cause bees.

Smoking increased in the 20th century, facilitated by mechanization. Cigarettes were no longer only for the rich or royal, and addiction and associated health difficulties increased. Smoking is still increasing, but since the 1990s some governments have taken a bold approach, removing branding and showing images of the diseases caused by smoking on packaging. Let's hope this indicates the future breaking of humanity's tobacco habit.

The tobacco plant is vulnerable to parasites, such as tobacco beetle and tobacco moth, which can cause severe damage to crops and consequently great financial loss.

The whole aconite plant is poisonous, but it does have uses in homeopathy.

Poison dart

ACONITE *ACONITUM NAPELLUS* (RANUNCULACEAE)
COMMON USE POISON

It is said that the Greek goddess Hecate, who presided over magic, spells and poisons, discovered this strikingly handsome plant. Aconite is believed to have grown in a region called Aconas or Akonas, near the ancient city of Herakleia, from which it gained its name. Another common name, 'monkshood', refers to the flower's iconic 'helmet'. *Napellus* comes from the root system, which is napiform, consisting of swollen, turnip-like roots.

Hunters used aconite juice on arrows to poison their prey, leading to other common names, among them *pardalianchus* (leopard killer), *theripponon* (brute killer) and *lycoctonon* (wolf killer). The last is the basis of another common name, 'wolfsbane'. The use of aconite in hunting suggests another possible origin of its name – the Greek word *akon*, meaning dart or javelin.

Aconitin, which occurs in large quantities in aconites, is the most poisonous of the alkaloids, requiring only a small dose to kill. Numbness spreads from mouth and throat to the rest of the body. The heart beats irregularly, blood pressure drops, the airways close and drowsiness sets in. Death follows shortly afterwards. Some believe that Alexander the Great, King of Macedonia, who died in 323 BCE, was poisoned using aconite, since accounts by Diodorus Siculus (1st century BCE) and Plutarch (1st century CE) – both considerably after the fact, admittedly – suggest that he suffered a short illness after drinking wine. Poisoning was relatively common from the time of the early Greeks to the end of the Middle Ages, whether among royalty removing political opponents or among the common people removing spouses and lovers.

In the early 16th century Pope Clement VII, who lived in great fear of being poisoned, ordered the first modern human trials of aconite. Two convicted criminals, who were foreigners and had no family, were chosen, and as they were brought in, prayers were spoken, supposedly to annul any wrongdoing on the part of the authorities. In front of a small audience of invited guests, aconite was given to each in a marzipan cake; one of them was then given an antidote created by the Pope's physician. The other prisoner died a horrible death. This spurred a movement of barbaric human trials to find antidotes for poisons to curb the fears of the rich and powerful.

Aconite has not been widely used since about that time, and its use stopped almost entirely in the 20th century. Some rural communities in Slovenia create home-made spirits from *A. napellus* and the related *A. tauricum*. In 1992 the Slovenian pharmacognosist (a scientist who studies plants as the source of drugs) Pavle Bohinc described how aconite was applied externally for arthritis, gout and rheumatism. However, in 2021 the President of Kyrgyzstan, Sadyr Japarov, recommended aconite as an effective treatment for COVID-19. This resulted in the hospitalization of four people with symptoms of poisoning.

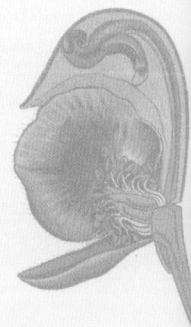

More than a weed

CANNABIS *CANNABIS SATIVA* **(CANNABACEAE)**
COMMON USES DRUG • BIOFUEL • COVER CROP

Cannabis is a controversial drug. Some believe it to be a valid medicine that can be used in the treatment of various illnesses. Others believe it is dangerous and mind-altering: a gateway drug to more serious substances. Yet the narcotic aspect of cannabis is only one part of its story. *Cannabis sativa* belongs to a relatively small family that also contains hops (p. 25) and hackberries (*Celtis* spp.). Cannabis is an annual herb and is dioecious, meaning each plant is either male or female. It comes originally from Central Asia: from northern Pakistan to Kazakhstan and neighbouring Xinjiang, China. It has since been introduced to every inhabited continent.

The origin of the genus name is unclear. It may come from the Thracian people, whose territory covered present-day Bulgaria and Turkey. The word may have been learned from ancient Semitic-speaking people, who would have come from the plant's native region. The name passed into the Persian language as *kanab*, then to Greek as *kánnabis*. Owing to the plant's long history of cultivation, widespread distribution and selection, there has been some confusion about the taxonomy of the genus. Some taxonomists believe there is only one widespread and variable species; others separate the plant into three: *C. sativa*, *C. indica* and *C. ruderalis*.

Broadly speaking, cannabis is split into three groups: plants cultivated for fibre; those grown to produce seed for hemp oil; and those used for medicinal or recreational reasons. They are separated owing to their chemical composition, specifically their proportions of phytocannabinoids, the substances uniquely found in cannabis. The most famous, and the one with the highest intoxicating effect – tetrahydrocannabinol or THC – is found in higher quantities in some strains of the plant, and these are commercially sold as marijuana. Strains with a lower volume of THC do not produce the same psychoactive effect, and are known as hemp. The English word 'canvas', a fabric traditionally made from hemp fibres, is thought to be derived from the Greek *kánnabis*. Over time, cultivars of cannabis have been selected to give higher yields of flowers in the marijuana strains and longer stems in the hemp. These cultivars are therefore distinct in their appearance and require different cultivation.

Cannabis has been grown since ancient times, and has been used for fibre, food, oil, manufacturing and religious purposes, as well as for medicine. It has been recorded at Neolithic sites dating back 12,000 years, making it one of the world's oldest domesticated plants. Several religions hold it sacred. Hindus and Buddhists have marked the use of its flowers and resins to induce a meditative state that eased communion with spirits, and in the Old Testament it is mentioned as a sacred oil. The first Chinese herbal, *Shennong Bencao Jing* (Shennong's Herbal), which dates back to 3000 BCE and is said to comprise the writings of the mythical emperor Shennong,

Cannabis is a short-day plant, so it starts flowering only
at the end of summer, when the nights become longer.

In Morocco, hairs from the cannabis leaf (known as trichomes) are called *kief*, and are chopped
and mixed with tobacco. This mixture is traditionally smoked using a pipe called a *sebsi*.

includes cannabis. Its recorded uses were for reducing tiredness and treating rheumatism and malaria. Cannabis also appears in the Egyptian Ebers Papyrus and on the Assyrian clay tablets excavated by Sir Austen Henry Layard in the 19th century (see p. 82). It was known to the Zoroastrians and recorded in their sacred book, the Avesta. In ancient Greece, the plant was used as a painkiller and to improve mood. Throughout medieval times in the Middle East, the herb was used for its psychoactive effects and medicinal uses, including in the treatment of epilepsy. Maimonides wrote in the 12th century that it was one of the most frequently used drugs. The legal banning of cannabis because of its psychoactive effects began only comparatively recently, however; in 1937 the United States banned it through the Marijuana Tax Act, and 1971 brought the comparable Misuse of Drugs Act in the United Kingdom.

Despite these bans, cannabis is the third most common drug in the world, after tobacco (p. 86) and alcohol. It is the oldest known fibre plant, and hemp fibre is now used for an array of applications, including in paper and textiles, and as a building material. Biofuel is also made using cannabis oil. Cannabis is used more and more as a 'cover crop', one that returns nitrogen to farmland soil during years when the soil lies fallow of other crops. Additional products are generated from the plant, such as animal feed. It is used to treat digestive problems in Nepal; farmers use it to feed livestock in Uganda; and it is taken as a sedative and tonic in northern Pakistan.

The Netherlands is often understood to epitomize unrestrained cannabis use, but in fact the country has never legalized cannabis production for anything other than medicinal use. The plant has, however, been decriminalized there since 1976, meaning that it is still illegal, but the legal system does not prosecute the individual for committing the crime. Uruguay was the first country to legalize cannabis production, in 2013, and Canada followed in 2018. By 2017 twenty-one American states had legalized medical cannabis, eight of them allowing its recreational use. Perhaps the legalization of cannabis will continue, or maybe drug laws will outlaw the plant again. The debate rages on.

The many uses of hemp fibre are becoming a focus once more. Hemp blocks are an innovative building material, advertised as natural bricks. They combine breathability and a low carbon footprint with the strength to be used in building houses. Having been used for ropes and sails during the age of sailing ships, hemp also produces tough fabric for the manufacture of clothes, and is environmentally friendlier than synthetic, non-recyclable materials. It is regarded as one of the strongest natural fibres in the world.

Fingerhuts and foxes' gloves

FOXGLOVE *DIGITALIS PURPUREA* (PLANTAGINACEAE)
COMMON USE MEDICINE

The foxglove is widespread in Europe, both in gardens and in nature. It is a biennial plant that creates a rosette of leaves the first year, and blooms during the second with a long spike of purple flowers speckled inside with brown. These markings guide bees to the centre of the flower so that they pick up pollen. The plant produces large quantities of very small brown seeds that germinate easily in a range of habitats and growing conditions. Many selections have been made for different colours and more abundantly flowering forms.

Digitalis comes from the Latin word *digitus* (finger). It was named in 1542 by the physician and botanist Leonhart Fuchs (who gave his name, incidentally, to the genus *Fuchsia*). Fuchs based the name on the German word *Fingerhut* (thimble), referring to the shape of the flowers. The origins of the common name 'foxglove' are harder to specify. It may be linked to an ancient musical instrument called the fox's glew, which had a similar shape to the flowers, or it may come from a fairy tale in which a fox wears flowers on its paws so that it can steal chickens without being heard.

Foxglove was first recorded in 1250, when it was prescribed by a Welsh family known as the Physicians of Myddfai, although it is not known how they discovered the plant or what it was said to cure. English and German herbalists began recommending foxglove tea as a diuretic in the 16th century, yet it was not until 1785 that its usefulness as heart medication was formally published.

The story goes that the 18th-century botanist and chemist William Withering – writer of *An Account of the Foxglove and Some of Its Medical Uses* (1785) – had learned from an old woman in his home county of Shropshire a herbal remedy for treating oedema, then known as dropsy (swollen feet and legs caused by a poorly functioning lymphatic system). After several years, Withering deduced that foxglove was the active ingredient in this remedy. He was asked to assist Erasmus Darwin (the grandfather of the celebrated naturalist Charles Darwin) in treating a dropsy patient, whose health subsequently improved. However, a few months later Erasmus's eldest son (also Charles), a medical student in Edinburgh, died before completing his doctoral dissertation. Grieving, Erasmus added the use of the foxglove treatment to his son's incomplete dissertation, crediting him with the discovery and making no mention of Withering. This began a bitter rivalry between Erasmus and Withering, and the two men never reconciled.

In England in 1930 digoxin, one of the primary compounds in digitalis, was first isolated from the related *D. lanata* by Dr Sydney Smith. Digoxin, which is a key part of many cardiac drugs, increases the force of the contractions of the heart. This means the heart can work harder with less oxygen. However, the dosage must be just right. If it is too high, it can cause arrhythmia, an abnormal heart rhythm, which can be fatal.

The individual flowers of the foxglove were believed
in folk tradition to be the homes of fairies.

Stinging nettle; see p. 116.

HEALING HERBS

Healing Herbs

Despite the astounding advances made in medicine over the last three centuries, many of the healing herbs of our ancestors are still used today. Beginning thousands of years ago, herbalists dedicated their lives to the study of these plants. They spent countless hours in nature, searching for species, noting where to find them and meticulously recording their healing properties. The correct identification of plants was a necessity, since mistakes could be deadly. The ancient Greek word *phármakon* – the origin of the word 'pharmaceutical' – meant both poison and cure.

The word 'drug' comes from the German *droge* (dry), since drying herbs was the first step in creating medicine. Once the herbs were harvested, they were taken to market to be sold, or were sent to physicians and healers. These medical practitioners furthered research into unrecognized properties and health benefits of the herbs. For centuries herbs were at the forefront of medicine and used in a variety of healing practices.

Information concerning how to identify herbs and their various healing qualities travelled alongside them. When the Roman army marched into other countries, they took plants with them. They carried yarrow (p. 100), for instance, to treat wounds suffered in battle. They recorded the medicinal and culinary use of herbs by the people of the countries they conquered, and traded information with locals about the herbs they travelled with. The Romans are thus considered responsible for both the spread of many healing herbs outside their native areas, and the naturalization of some species.

Meanwhile, the written works of early Greek and Roman scholars moved eastwards to Asia. Middle Eastern physicians learned from them and from the established tradition of Ayurvedic medicine practised in India. This resulted in a coalescing of the progress made in Eastern and Western medicine. The Persian polymath Ibn Sīnā (commonly Latinized to Avicenna) is widely regarded as the writer who amalgamated these writings most authoritatively into a single work: *The Canon of Medicine*, published in 1025.

In the 17th century the herbalist Nicholas Culpeper aimed to enable everyday English people to make their own healing remedies without having to buy expensive foreign herbals. It was his ambition to produce a cheap herbal detailing easily found and mostly native herbs, and in 1652 he published the comprehensive book *The English Physitian*, later known as *Culpeper's Complete Herbal*. Scientific advances made during the years of the Industrial Revolution produced alternative medicines, such as mercury, which, despite their dangers, doctors favoured over herbal remedies. In the 20th century the rise of 'modern' medicine (especially the development of antibiotics) came at the expense of traditional and herbal medicine, which is now regarded with scepticism. Pharmaceuticals developed by multinational companies are often preferred to herbal remedies and discussed as if they are entirely unrelated.

Despite a more scientific, technology-based approach in so-called developed countries, herbs are still the basis of many medications. Meadowsweet (p. 108), for example, is used in the production of aspirin. St John's wort and garlic (pp. 103, 120) have many recorded traditional uses and potential applications in modern medicine.

We now have the luxury of going back in time through herbals and pharmacopoeia to test the validity of historical claims. While some uses are fanciful or were clearly wrong, many applications discovered by our ancestors are correct and have been proved so again and again in rigorous tests.

This chapter examines the iconic healing herbs that have been called on throughout time: species that have been used around the world to heal sickness, and plants that are still commonly found today. The text notes the most recent scientific developments, whether new healing qualities that are currently under investigation or results that have been scientifically proven.

Achilles herb

YARROW *ACHILLEA MILLEFOLIUM* **(ASTERACEAE)**
COMMON USES COAGULANT · WOUNDS

In many parts of the world yarrow is a common roadside plant. This member of the daisy family grows in short grassland and arable land in the temperate regions of the northern hemisphere, often in abundance. Its deeply divided, feathery, fern-like leaves – to which the species name, *millefolium* (thousand leaves), refers – make it easy to identify.

Yarrow may be one of the oldest herbs known to have been used by humans. Pollen from an *Achillea* species was found at a Neanderthal grave in Shanidar Cave, an archaeological site in the Zagros Mountains, Iraq. The pollen was analysed and discovered to be 65,000 years old. One of the bodies in the cave appears to have been buried on a bed of the shrub *Ephedra*, and flowering plants including *Achillea* were placed on top as a bouquet.

The word *achillea* comes from the mythical Greek warrior Achilles, who applied yarrow leaves to the wounds of his fellow soldiers during the Trojan War to slow blood loss. In the 1st century CE Dioscorides recorded a range of blood-staunching uses for the plant, as well as its capacity to reduce inflammation and the symptoms of dysentery. At about the same time Pliny argued its usefulness in easing digestive problems, bleeding, earache and menstrual discomfort.

Many of yarrow's folk uses in British herbals of the Middle Ages were also linked to blood. It was called 'blood wort' and

'soldier's woundwort', and herbals noted its ability to stop bleeding – particularly nosebleeds – and to help with high blood pressure. In the 17th century John Parkinson wrote that the plant had even gained the nickname 'nosebleed', being used so regularly for this problem.

Yarrow's medical properties are recognized today. In Germany and Italy, it is used primarily for digestive troubles, but in Hungary it is still used to treat wounds. Some Indigenous people of North America use yarrow for wounds, bleeding and infections, and to treat skin afflictions. The late American herbalist Steven Foster called it a 'herbal Band-Aid', and indeed scientific studies have shown that yarrow has significant wound-healing and blood-stopping properties. Further research has shown positive results for the herb's antiparasitic, anti-inflammatory and antioxidant qualities. Yarrow is not widely used in present-day medicine – by pharmaceutical companies, for instance – because it causes contact dermatitis in some individuals. These people are particularly sensitive to the chemical compounds guaianolides, which appear in high concentrations in *A. millefolium* and some closely related species.

Yarrow has become a common roadside weed around the world, because it is adaptable and spreads quickly by self-seeding.

In French, St John's wort is known as *chasse-diable* (devil chaser),
from the traditional belief that it can ward off evil spirits.

Herb of St John

ST JOHN'S WORT *HYPERICUM PERFORATUM* (HYPERICACEAE)
COMMON USES ANTIDEPRESSANT • HIV

St John's wort is a widespread Eurasian species that has spread outside its natural area and become an invasive weed in North America and temperate regions of the southern hemisphere. Its genus, *Hypericum*, contains more than 700 species, which span the globe; they range from annuals to herbaceous perennials and shrubs, and almost all have bright yellow, star-shaped flowers. The name of this particular species, *perforatum*, comes from the cut appearance of the leaves.

St John's wort takes its name from John the Baptist, the preacher who is thought to have baptised Jesus. He was imprisoned and subsequently beheaded for publicly criticizing King Herod. From the 11th century onwards, on the anniversary of his death, St John's wort was traditionally hung in Catholic homes in Europe. This commemorative use probably gave rise to its name, since the Latin *hypericum* derives from the Greek *hyper* (above) and *eikon* (image).

Early mentions of the plant are made by ancient Greek scholars and herbalists. Dioscorides recommended it for healing burns and for pain relief; the 2nd-century surgeon and philosopher Galen used it for snake bites, depression and melancholy; and Pliny mentioned it as a diuretic and, again, a treatment for burns. By the Middle Ages, St John's wort had become a symbol of purity, and was used to ward off evil spirits. The Swiss physician Paracelsus wrote in the 16th century of its use in amulets to protect the wearer against spells

and curses. According to the first edition of the London Herbal (*Pharmacopoeia Londinensis*) in 1618, the oils of the plant were used to make a tincture to treat external wounds and bruises, and it was added to wine to cure internal ailments, such as swelling and inflammation.

Species of *Hypericum* contain a chemical compound called hypericin. In 1988 researchers at New York University and the Weizmann Institute of Science in Israel considered it a potential revolutionary treatment for HIV (human immunodeficiency virus), which causes AIDS (acquired immune deficiency syndrome). A later study by the two institutions reported that the hypericin treatment effectively rendered the HIV cells inactive. Hypericin has shown similar preliminary results on leukaemia.

Consequently, St John's wort has been one of the most intensively investigated medical herbs in the world since the turn of the millennium. Yet its function as an antidepressant remains the only medicinal use that is so far accepted by the European Medicines Agency. Research has revealed that hypericin could be dangerous for livestock, and fears that St John's wort could have similarly damaging effects on people have altered the perception of the plant amid otherwise positive results. Its proponents emphasize the 2,000 years of its use by humans and point to a multitude of possible medicinal uses that require further research.

Self-healing herb

SELF-HEAL *PRUNELLA VULGARIS* (LAMIACEAE)
COMMON USES COAGULANT • TEA

Prunella vulgaris, as its species name suggests, is a very common plant and can grow in a range of soils and habitats. It is found in both temperate and subtropical climates in much of the northern hemisphere, and has been introduced into South America and parts of Oceania. Its ability to take over waterlogged lawns and grasslands has led to its reputation as a weed. However, if grown in very dry conditions, it does not spread so quickly.

Self-heal was first recorded as a treatment by early German physicians, who used it for sore throats. The German name for a throat inflammation, *die Braune*, was corrupted in English to 'brunella' – hence the Latin classification, *Prunella*. The name 'self-heal' was described in the 17th century by Nicholas Culpeper: 'When you are hurt, you may heal yourself.' Some decades before, John Gerard had boldly stated that there was 'no better wound herb' than self-heal. The herb contains saponins – chemical compounds common in herbs – which are known to help stem bleeding by encouraging the blood cells to coagulate.

Self-heal is commonly given to livestock as a rich and nutritious food, and is used in veterinary medicine to cure small injuries. A study conducted in British Columbia in 2007 recorded positive results concerning self-heal's use on organic farms as a treatment for wounds, and advocated that more farms use the herb instead of synthetic medicinal products.

Yet this herb has still other medicinal properties that are beneficial to humans. In South Korea it is used to treat dermatitis and skin allergies. In Unani medicine, a traditional approach followed in Muslim cultures in South and Central Asia, self-heal eases sore throats, colds and headaches. In Kashmir it is boiled and the steam inhaled to relieve headaches, and it features in Chinese traditional medicine in the treatment of hepatitis, gonorrhoea and tuberculosis, as well as in herbal teas. It is the main ingredient in Guangdong tea, popular in southern China as a calming infusion.

Self-heal's anti-inflammatory qualities are supported by its high levels of rosmarinic acid (although the concentration varies according to the genetic make-up of the plant, and the climate and location in which it grows). This chemical is found in other plant species commonly used in Asia to treat inflammation, such as *Clematis mandshurica* and *Trichosanthes kirilowii*. In addition, studies in the coming years are expected to develop understanding of self-heal's anti-cancer and antioxidant properties.

Self-heal is a key ingredient in *wong lo kat*,
one of China's most popular tisanes.

The bitter taste of feverfew means that the plant has only
a few pollinators; others are repelled by its strong scent.

Fewer fevers

FEVERFEW *TANACETUM PARTHENIUM* **(ASTERACEAE)**
COMMON USES PAINKILLER • FEVER

With a name like feverfew, one would expect this plant to have a long history of fever medication. While it is best known for that, the common name in fact comes from a mistranslation or misspelling. During the Middle Ages, the plant was called 'featherfoil' because of the feathery edges of its leaves. After some time, this became 'feverfoil' and then, as we know it, 'feverfew'. The name was solidified in the Middle Ages as the treatment of fever became the herb's primary use.

At this time the herb – a perennial with attractive white, daisy-like flowers – was planted around homes. Malaria, a serious and sometimes fatal disease commonly spread by mosquitos, was widespread in Europe and plagued many people. It was originally believed to be caused by poor air – the name comes from the Italian *mala* (foul) and *aria* (air) – and the strong-smelling feverfew was believed to purify the air and ward off the disease. This may inadvertently have worked because the herb's strong aroma repelled the mosquitos that carried the disease. In 1633 feverfew was replaced in the treatment of malaria by the more effective cinchona bark (*Cinchona officinalis*), most commonly known as quinine. This famous 'fever tree' rendered other fever-related herbs unnecessary.

John Parkinson recommended feverfew as 'very effectual for all pains of the head', and a little later, in 1772, the herbalist John Hill wrote that 'in the worst headache, this herb exceeds whatever else is known.' But this knowledge and use of feverfew diminished too – until the 1970s, that is. When the wife of the head of Britain's National Coal Board suffered from regular migraines, a miner recommended chewing feverfew leaves, typically two leaves per dose. She experienced immediate relief and was freed from migraines after just over a year. A story detailing her cure was published, and the use of feverfew in treating migraines gained publicity. This was substantiated by a trial carried out in London, in which individuals who took feverfew regularly had notably fewer migraines than those who did not.

The Canadian health service has approved feverfew specifically for migraine medication. Elsewhere in the Americas, however, the herb is used more extensively. The plant is taken for earache in Venezuela, for menstrual complaints in Mexico and for poor digestion in Costa Rica. The Kallawaya people of the Bolivian Andes use feverfew to treat stomach-related illnesses, such as kidney pain, stomach ache and colic. A study carried out in 2011 advocates for the herb as a 'safe, highly important, medicinal plant for general mankind'. Yet most medical practitioners maintain that more scientific study is required before it can become a mainstream medicine.

Queen of the meadow

MEADOWSWEET *FILIPENDULA ULMARIA* (ROSACEAE)
COMMON USES DRINK · MEDICINE · FLAVOURING

This eye-catching herb can be found throughout Europe and western Asia, as well as in North America, where it is now common. Meadowsweet is a tall plant, growing to 1–2m (3–6½ft) and dominating wet meadows, but it is also tolerant of drier spots and summer drought, making it a popular garden plant. It is a member of the rose family, which also contains apple, pear, cherry, almond and many other useful plants.

The genus name *Filipendula* is from the Latin for 'hanging thread', because of the fine plumes of white flowers that catch the wind to carry the seeds away to a new spot. The species name *ulmaria* refers to the leaves, which resemble those of the elm tree (*Ulmus* spp.). The common name indicates its use to sweeten the honey-based alcoholic drink mead, a staple during the Middle Ages in Europe, and many European countries have a similar name for it; in Germany, for example, it is *echte Mädesüß*. When mixed with spices or herbs, mead was called *metheglin* (roughly 'healing liquor').

Archaeological remains found in Fife, southeastern Scotland, and dating back to the Bronze Age revealed bodies buried with the flowers of meadowsweet, indicating that it may have been used not only for its herbal qualities, but also for symbolic reasons. Additionally, dye was made from the roots and foliage, and perfumes from the flower buds. Another common name for meadowsweet, bridewort, originates in its common use in the Middle Ages to line the path of the bride in church, or to make decorative garlands.

The foliage of meadowsweet has been used as a remedy for a list of ailments, among them fever, gout, diarrhoea, heartburn, colds and rheumatism. One of its key components is salicylic acid, which almost everyone has taken even without knowing it. It was first isolated in plants of *F. ulmaria* (then *Spiraea ulmaria*) to make aspirin (a name that is a combination of *Spiraea* and 'acetylation'). However, the willow tree (*Salix* spp.) has taken all the credit, largely because it is safer and has fewer adverse effects than meadowsweet.

All parts of the herb have a subtle flavour, and this has made it an important food and flavouring ingredient. The whole plant is edible once cooked, but the flowers are most commonly used, since the leaves taste a little more bitter than the flowers. It makes a delicious herbal tea and is popular among foragers, since it grows so plentifully.

Meadowsweet's anti-inflammatory properties are being investigated in the
hope that it will be effective against irritable bowel syndrome.

Milk thistle is a popular herb and has many uses
associated with its function as a detoxifier.

Hangover herb

MILK THISTLE *SILYBUM MARIANUM* (ASTERACEAE)
COMMON USE CANCER

Milk thistle, a member of the daisy family, was originally native to Asia and southern Europe, but is now widespread around the world. In fact, it has become a noxious weed in South America, Africa, Australia and the Middle East, presenting a great problem on farmland, since its high concentration of nitrates can make it toxic to cattle and sheep.

This tall herb can grow to 2m (6½ft) or more, and grows as an annual or biennial, producing large purple flower heads with spiky leaves. According to legend, the white of the leaf veins comes from a fallen drop of the Virgin Mary's milk. This tale has led to several of its common names, such as 'Mary thistle' and 'St Mary's thistle', and is the origin of the species name *marianum*. It was traditionally used as a galactagogue – a herb that stimulates milk production in humans and other mammals – hence the common name 'milk thistle'.

The herb is known to have been used by the ancient Greeks and Romans. It was mentioned in the 1st century CE by Dioscorides, who recommended it to treat snake bites, and at about the same time Pliny wrote that it could be mixed with honey to help digestion. By the Middle Ages, milk thistle was well known for its treatment of liver illnesses and its help with the elimination of toxins. John Gerard hailed it as 'the best remedy that grows against all melancholy diseases', and European settlers introduced it to North America, where their physicians used it to treat

gallstones and discomfort associated with the liver and kidneys, and with pregnancy.

Milk thistle contains silymarin, a chemical with unique hepatoprotective (liver-protecting) qualities, and is thus effective for the treatment of liver disorders, including those involving damage caused by the excessive consumption of alcohol or other drugs. The herb is also a powerful antioxidant, and has been reported to reduce chronic inflammation and treat diabetes, cardiovascular diseases and infections. Clinical studies have shown that it is also useful as a treatment for diabetes caused by alcohol-induced cirrhosis of the liver.

Milk thistle is currently one of the most popular herbal products in the United States. It is commonly found as capsules or tablets, rather than as a tea (as it has more traditionally been used); it is accessible and has no recorded safety concerns. It is used as a detoxifier after chemotherapy, and its role in cancer treatment may grow, depending on the results of further research. Investigations carried out in North America into the herb's usefulness in treating kidney problems have been positive, so it is likely that the range of applications – both as a herbal remedy and as a specialist medicine – will expand.

The flower to help you breathe

ELECAMPANE *INULA HELENIUM* (ASTERACEAE)
COMMON USES RESPIRATORY PROBLEMS • ABSINTHE

With its flowery 'faces' looking up at the sky, this lovely sunflower-like herb is a striking plant in the medicinal herb garden. It is very tall – over 1.5m (4½ft) – and has a rosette of large leaves growing from a big, fleshy root. Upright flower stems emerge in the summer to bear its yellow, daisy-like flowers.

The Latin name, *Inula helenium*, comes from Helen of Troy, daughter of the Greek god Zeus. She is said to have carried a bunch of these flowers when Paris stole her away; an alternative explanation is that the plant sprang from her tears as she was carried off. As with all myths, there are many versions of the story. 'Elecampane' is thought to come from the Middle English name for the herb, *elena campagna*, which comes from the same root as the Latin name. Other common names, which allude to the plant's use in herbal medicine, include scabwort, horseheal, elf dock and wild sunflower.

Inula is native to western Asia and is thought to have been naturalized in Europe and southern Britain since the Bronze Age. It is popular with gardeners for its architectural form, long summer flowering and herbal interest. It can be propagated easily by digging up the fleshy roots and cutting them into smaller pieces for replanting.

The root is harvested for making herbal treatments. The Greek physician Hippocrates wrote in the 4th century BCE about the virtues of using elecampane as a cure for disease of the hip. The Romans made a candied confection of the root to aid digestion and improve temper. This sweetness was heartily enjoyed much later, in the Elizabethan court in England, where it was made into cakes that were eaten to sweeten the breath.

The mysterious 12th-century Italian healer Trota, from the Salerna school of medicine, advised that the whole plant be cooked in wine and oil and placed on the belly to cure colic. John Gerard and Nicholas Culpeper, in the 16th and 17th centuries respectively, recorded elecampane's value in a compress to relieve gout and sciatica. Apothecaries later began distilling the volatile oils to make a camphor-like treatment for such respiratory conditions as bronchitis and wheezing. It seems there is no end to the herb's usefulness; the French and Swiss even added it to absinthe, their famous alcoholic spirit (see p. 74).

A more light-hearted application is offered by the English gardener Stephen Blake. In *The Compleat Gardeners Practice* (1664), he wrote: 'To be revenged on a person who steals your tulips, sprinkle dry powdered elecampane root on clove gillyflowers [carnations or pinks], give to the party who will delight to smell it and … they will fall sneezing until tears run down their thighs.'

Herbalists today use elecampane as a tincture for the treatment of respiratory problems and in tea to help with digestion. As always, it comes with a health warning; some people have allergic reactions to plants in the daisy family and should avoid using preparations made from this herb.

Inula is a hardy perennial that does well on rich, damp soil and in partial
shade, where it will flower from June until September.

The flowers of heartsease, with their familiar bicoloured 'faces', have in recent
years become very popular as a decorative garnish for food.

Potions for the heart

HEARTSEASE *VIOLA TRICOLOR* (VIOLACEAE)
COMMON USES ANTI-INFLAMMATORY • SKIN DISORDERS

Many people are familiar with violas as bedding plants – annuals grown en masse for sale in garden centres and supermarkets. Yet these plants reflect only a small part of the genus. *Viola* belongs to a large family of about 1,000 species, most of which are violas. Their typical flowers are easily recognized, even considering the variability of leaves, habitats and growing conditions.

V. tricolor, an annual or short-lived perennial with creeping roots, is a common wild flower in European meadows. It is also found naturally in western Asia, and was introduced to North America by early colonists. As the species name suggests, the flowers are made up of three colours: the top two petals typically purple, the middle two white, and the lower yellow. As well as attracting pollinators, the three-coloured petals were considered symbolic of the Christian Holy Trinity.

According to another tale, a viola bloom was struck mistakenly by one of Cupid's arrows and bruised purple and yellow. The juice from the flower could thereafter form a love potion. Violas were indeed a key ingredient in love charms, and as a result the plant had a range of love-related common names, including 'love-in-idleness', 'cuddle-me', 'call-me-to-you' and 'love idol'. The first name is used by Oberon in Shakespeare's *A Midsummer Night's Dream* (II:i): 'Yet marked I where the bolt of Cupid fell./ It fell upon a little western flower/ Before, milk-white, now purple with love's wound,/

and maidens call it "love-in-idleness".' It was also given the Anglo-Saxon name 'banwort' or 'bonewort' for its healing properties.

Heartsease has been used as a medicinal plant in Europe for centuries. Infusions made from its leaves were said to aid digestion, and it was documented in European herbals as a remedy for inflammation, asthma, eczema (for which it is also used in North America) and other skin diseases. It came into the spotlight in the early 19th century, when an article in the *Medical Journal* reported it as a valuable remedy for such skin disorders as scald-head (a disease of the scalp) and ringworm. Heartsease contains the flavonoid rutin and also salicylic acid, both of which are reported to reduce internal inflammation and skin swelling. In Lithuania, the herb is used for cuts and in tonics, while in the United States, it is used mostly to treat sore throats and colds.

The flowers have in recent years become very popular in kitchens as a garnish for salads and desserts. The leaves are also used to make a tea, and, being mucilaginous (sticky, in the same way as okra), to thicken soups.

Heartsease contains high concentrations of flavonoids, chemicals found only in the plant kingdom and which have shown great promise in the treatment of such serious illnesses as AIDS, Alzheimer's and diabetes. The herb also contains high volumes of cytotoxic chemicals, which may prove useful in stopping the spread of cancerous cells – but further research is needed.

From sheets to soup

STINGING NETTLE *URTICA DIOICA* **(URTICACEAE)**
COMMON USES CULINARY · TEXTILE

Whether by accidentally grasping it or brushing against it, most people know the piercing sting of a nettle. The leaves are covered in small hairs called trichomes, which, when they come into contact with skin, act like needles. They release chemicals into the skin, sometimes with an accompanying sharp pain. A rash, normally of light-coloured bumps, appears on the surface of the skin, and the burning and often itchy feeling can last for up to half an hour. Allergic reactions are rare, but can occur and are life-threatening. This uncomfortable attribute has given the plant a negative image as a weed.

Not all species in the genus *Urtica* can sting, while others are significantly more painful than the common stinging nettle. The effects of *U. spatulata*, which is found in Java, can last for up to a year. The genus *Urtica* is a widespread one, found in much of the northern hemisphere and in southern Africa, Australia and the Andes mountains of South America. It grows underground structures called rhizomes, which spread quickly so that the plant forms large groups.

The name of this species, *dioica*, refers to the fact that it is dioecious, meaning that individuals are either male or female, and that consequently both are required for fertilization and the production of seed to occur. These flowering plants grow up to 2m (6½ft) tall and the seeds are dispersed by wind, so they do not require the exciting or attractive flowers that would attract pollinators. *Urtica* comes from the Latin

urere (to burn), and 'nettle' comes from the Anglo-Saxon word *netel* (needle). It might seem obvious that the latter etymology comes from the plant's sharp sting, but it could also be related to the herb's use as a textile.

The oldest record of textiles made from stinging-nettle fibres was found in Denmark, where remains of the plant were discovered and dated to the Bronze Age. This suggests that nettles played a much more important role in the early textile industries than had previously been realized; the focus has always been on the long history of flax and hemp for that purpose (see cannabis; p. 90). Later, in the Iron Age, linen was introduced, phasing out the production of nettle fibres.

However, nettle fibres were used in Scotland in the 16th and 17th centuries to make a fibre similar to that produced from hemp and flax. The early 19th-century poet Thomas Campbell wrote: 'In Scotland, I have eaten nettles, I have slept in nettle sheets, and I have dined off a nettle tablecloth. The stalks of the old nettle are as good as flax for making cloth. I have heard my mother say that she thought nettle cloth was more durable than any other species of linen.' During World War I, Germany ran out of cotton and instead had to use nettle in its textile-manufacturing industry. Inevitably, the plant's stinging hairs make it more difficult to harvest than flax and hemp, so it has not become commercially viable except in very small-scale operations.

Nettles also have a long history of use as medicine. The ancient Egyptians employed

Nettle leaves are rich in vitamins and minerals,
but they are a sadly underappreciated source of food.

Nettle is at its most tender when the fresh new shoots are about
the length of a finger; as they mature, they become more fibrous.

the stinging nettle in an infusion to treat arthritis by warming the joints and making them less painful. This process, known as urtication, is still a common folk treatment for this painful condition. Hippocrates wrote that the herb could be used as a treatment for snake bites and scorpion stings, as well as an antidote to poison. In 200 CE, Galen listed a range of uses for the stinging nettle, from a diuretic to a treatment for nosebleeds and dog bites. Another species, *U. pilulifera* (Roman needle), is known to have been used by Roman soldiers to treat tired legs after long marches.

Similarly, nettles were used as a tonic to increase strength for agricultural work in the Middle Ages. Hildegard of Bingen recommended the seeds in the 12th century to soothe stomach ache, and Nicholas Culpeper wrote in 1653 of mixing the herb with honey to soothe mouth infections and relieve wheezing. He also recommended the process of urtication for arthritis, as well as for gout and sciatica. In 1737 Elizabeth Blackwell wrote that the stinging nettle could be taken to aid internal bleeding, while the juice of the plant could be applied directly to staunch wounds and stop nosebleeds.

In the 20th century the nettle gained a reputation as a nutritious food, and nettle beer was consumed until the 1930s as a well-known remedy for rheumatism and aches. An infusion of the fresh leaves is also soothing and healing as a lotion for burns.

Today, the stinging nettle is increasingly becoming a popular culinary herb. The fresh shoots are harvested in the spring, when the stinging chemicals are in lower concentrations, and the leaves are washed thoroughly. They are made into nettle tea or soup, or can be cooked and served as a green vegetable (all stinging effects disappear once the leaves wilt under heat).

At present, there are reports from every inhabited continent on the various medicinal uses of stinging nettle: treating bruises in Guatemala and Cuba, asthma in Brazil and Mexico, and arthritis in Germany, the United States and Peru. Recent studies of the roots of the nettle have shown them to reduce benign prostatic hyperplasia (BPH), a condition that affects a large percentage of men over the age of 50, enlarging the prostate and causing discomfort. Perhaps, then, this plant so strongly associated with inflicting pain may become increasingly appreciated for alleviating it.

The wonder herb

GARLIC *ALLIUM SATIVUM* **(AMARYLLIDACEAE)**
COMMON USES FOOD • ANTIBACTERIAL

Garlic is widely recognized as a culinary herb, but it has also historically been considered a wonder drug for its multitude of medicinal uses. It is regarded as the second-oldest herb in the world, after the Chinese *ma huang* (*Ephedra sinica*). *Allium* is a very large genus of about 1,000 species, also containing other important food species, such as onion, chive and leek. Apart from *A. synnotii*, which is found in South Africa, they are almost entirely plants of the northern hemisphere. Various species in the genus are ornamental and are therefore popular with gardeners for their dramatic flower heads and popularity with pollinating insects.

Remains of garlic have been found during archaeological investigations in caves dating from 8,000 BCE, meaning that early hunter-gatherers would have used it. Sumerian clay tablets from around 3,000 BCE are the first true written source of garlic's use. The herb was also used in ancient China and Japan, primarily as a preservative, and was often served with the meat it preserved. There is evidence that it was also used to treat sadness and depression. The Sanskrit text *Charaka-Samhita*, an ancient Indian herbal, recorded its use in the treatment of heart disease and arthritis.

Garlic was arguably the most important herb in ancient Egypt, and an impressive 22 uses for it are recorded in the Ebers Papyrus. The Egyptians pledged oaths over garlic as people do today with the Bible or the Qur'an, and it has been found in important temples and places of worship. It was buried with the dead, including the young pharaoh Tutankhamen, who died in 1323 BCE. The Egyptians believed that garlic (and onions) could increase an individual's endurance and strength, and garlic was reportedly part of the daily rations of the enslaved people who built the pyramids. More prosaically, they made a tincture of garlic, vinegar and water for use as a mouthwash and to treat toothache; this is probably why the Egyptians were called 'the stinking ones' by the Greeks.

By the time of the ancient Greeks and Romans, the use of garlic had expanded. In the 1st century CE Pliny recorded more than 60 different applications, including the treatment of snake bites and infections, and the general detoxification of the body. Garlic also clearly had symbolic significance, since it has been found at sacred grounds, including the ancient Minoan temple of Knossos on the island of Crete. In Homer's *Odyssey*, Ulysses uses garlic to fight off the evil sorceress Circe.

The Romans spread the plant through the European continent, and during the Middle Ages it became known as a charm against evil. It was hung over beds to fend off witches, placed underneath pillows to protect the sleeper from bad dreams, and set above doors to bar evil from entering. The belief that it could be used as protection from vampires is thought to have arisen in Eastern Europe during the Middle Ages, was immortalized in Bram Stoker's *Dracula* (1897), and has been part of popular culture ever since.

Take care not to bruise garlic when planting,
since damaged cloves are prone to mould.

Legend has it that during the construction of the Egyptian pyramids, there was a shortage of garlic, so workers' rations of it were cut – but the enslaved workers refused to continue without this energy-boosting food.

Hildegard of Bingen and the teachings of the medieval Salerno medical school both note the use of garlic, which was believed to give protection during plagues thanks to its strong scent. Garlic was one of the most prized plants in the early physic gardens, which were created at the end of the Middle Ages to discover and record the medicinal qualities of herbs, and to investigate new medicinal plants.

In the 17th century Nicholas Culpeper called garlic a 'remedy for all diseases and hurts'. European explorers took it to the Americas, where Indigenous peoples adapted it to their culture and used it in tea. The herb grew quickly in popularity in North America and was featured in many health books and herbals from the 18th to the 20th century. During World War I, garlic juice was essential in the treatment of battle wounds to avoid infection.

Garlic is a treasured ingredient in South Korea, and is found throughout the culture of that country. According to folklore, Hwanung, the founder of the first Korean kingdom, gave a bear and a tiger 20 bulbs of garlic. He promised that if they stayed out of sunlight and ate only garlic and mugwort (*Artemisia vulgaris*), he would make them human. The bear succeeded and was transformed into a woman. Garlic features in a wide range of traditional dishes, including Korean fried chicken, and the flower heads are made into the popular pickle *manul chang-achi*. South Koreans ate an average of 7.6kg (just over 16½lb) of garlic per person per year between 2011 and 2021. The only Koreans who do not eat garlic are Buddhist monks, who believe that this strong-smelling herb can increase anger and other unwanted desires.

The question is whether garlic is used to its full potential today. A great deal of it is consumed around the world as a culinary product. The Royal Botanic Gardens Kew points to garlic's antibacterial and antimicrobial compounds, which are good for overall health, and to its well-documented beneficial effects on the respiratory system. Therefore, the consumption of garlic as a food has a positive effect on health. Yet its medicinal qualities are often set aside in favour of those of other herbs or 'scientific' medicine.

The answer, then, is that we are probably not exploiting it to the full. The renewed interest in garlic over the past few decades may give it a greater role in the future.

Breaching barriers

BEAR'S BREECHES *ACANTHUS MOLLIS* (ACANTHACEAE)
COMMON USES ANTI-INFLAMMATORY • ORNAMENTAL

Bear's breeches is known as a tough, easy-growing perennial that survives in dry landscapes in its native Mediterranean, but is tolerant of a range of soils, making it a popular garden plant. It has unfortunately become invasive in some places, particularly in New Zealand, where it outcompetes local species.

The plant forms a rosette of spiky green leaves – hence its Latin name *Acanthus*, from the ancient Greek *akanthos* or *akantha* (thorn or spine) – from which a flower stalk emerges. The origin of the common name 'bear's breeches' went largely unquestioned until the botanist William Stearn investigated its meaning in 1996. The word 'bear' probably comes from the German *Barenklaw* (or variants) and refers to the similarity of the pointed bract (essentially a leaf structure that looks like a petal) to the claws of a bear. One possible reason for 'breeches' is the fact that the 16th-century herbalist John Gerard first mentioned the plant under the name 'beares breech of the garden'. He may have meant to refer to the plant's reputation for escaping ('breaching') its space and encroaching on that of other plants.

The leaves are very attractive and can be found on architecture. The defining characteristic of Corinthian columns (the third of the three classical orders, and found on ancient Greek temples) is the ornate patterns of acanthus leaves, and the foliage is also visible on Persian architecture from the 4th century BCE. The true significance of bear's breeches in this context is not fully understood, but it seems to have been of some religious value to the Persians, since its likeness can be found on sacred artefacts. There are few recordings of the ancient medicinal uses of the plant, although Dioscorides did recommend the roots in the 1st century CE to treat burns and to ease the pain of dislocated joints. It was also used as a diuretic, and was believed to soothe spasms and decrease flatulence.

In southern Europe acanthus has been used in traditional medicine to treat inflammation, toothache and intestinal problems. The leaves were crushed and applied to wounds and burns, and, more generally, the herb was used in baths to soothe aches and pains, and relieve headaches and swelling. In 2015 Portuguese researchers recorded the use of the seeds to treat haemorrhoids, diarrhoea, insect bites and digestive irritation. In Italy, the leaves are used to treat the debilitating skin condition psoriasis.

Despite these traditional folk uses, the plant has been studied little, so it is possible that there are still undiscovered important medicinal uses. Recent studies have produced positive findings. Research carried out in 2015 found high levels of the chemical compound hydroxyeicosatetraenoic acid, which reduces inflammation, and a study in 2017 showed antifungal properties (against the fungus *Candida*). The indications are that bear's breeches has great potential in medicine.

Bear's breeches is an iconic herbaceous plant for mixed borders, and
a tough species suitable for challenging situations in the garden.

Lemon balm was once called *melissophyllum*, which translates as bee leaf.

Bees please

LEMON BALM *MELISSA OFFICINALIS* **(LAMIACEAE)**
COMMON USES PHARMACEUTICAL • POLLINATORS

The original distribution of *Melissa officinalis* was across the Mediterranean and through Central Asia. It belongs to the mint and dead-nettle family, and its leaves have a strong lemony scent, leading to its popularity as a garden plant. It is excellent for pollinating insects, and its small white flowers are particularly attractive to bees; this has been shown to be because the chemicals it produces match the bees' chemical receptors. This once earned it the name 'bee balm', which is now used for its North American relation *Monarda*.

The Latin *melissa* is from the Greek for 'bee', which in turn is derived from *méli* (honey). The ancient Greeks planted lemon balm around the temple of Artemis, the goddess of wild animals and nature, to please bees and other pollinators. This practice of planting lemon balm to attract bees is still popular. In Bulgaria the plants are placed near vacant hives to entice bees inside. Lemon balm is often used as a companion plant, whereby it is planted beside flowering vegetables or fruit to increase the yield by ensuring good pollination.

The first record of lemon balm's use as medicine was by the ancient Greeks. Dioscorides wrote in the 1st century CE that it could treat wounds and could be mixed with wine to treat oral ailments. The word 'balm' refers to a healing aromatic. Traditionally used as a soothing and relaxing tonic, the plant spread across Europe and Central Asia towards the end of the Middle Ages.

During the first Elizabethan era in England, in the second half of the 16th century, the flowers of lemon balm were used in strong-smelling floral bouquets called tussie-mussies. At a time of widespread illiteracy, such bouquets were a popular method of expressing feelings and sending messages, and the strong fragrance served to overpower the evidence of poor sanitary conditions.

Today, lemon balm has been adopted by the pharmaceutical industry and is used regularly in soaps, shampoos, oils and creams. Its sedative and relaxing effects make it particularly desirable for use in bath and massage oils, or as a tea that is drunk before bed. Ingesting the plant is said to improve mood and cognitive ability, although the folk uses of lemon balm to aid memory remain difficult to substantiate. Recent research has shown a minimal improvement when the plant is applied as a balm to individuals suffering from dementia, but more research must be conducted to prove its usefulness in this regard.

Chilli pepper; see p. 147.

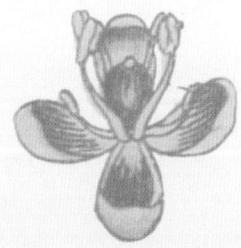

INDIGENOUS HERBS

Indigenous Herbs

People have since time immemorial explored their native areas to find herbs that could be used as food, drink or medicine, or that might have religious or cultural importance. Indigenous people across the world have formed bonds with the plants they live alongside. Indigenous herbs are those that originate in a particular area, and Indigenous people are defined as individuals who have inhabited an area since its earliest times, or who inhabited an area before the arrival of colonists. The understanding that Indigenous people have of their native herbs has been accumulated over hundreds or, in some cases, thousands of years.

The aim is always to harvest species respectfully. This means a sustainable and (normally) small-scale method that avoids abusing the land by overharvesting the precious natural resource. However, some Indigenous people have themselves been exploited for their knowledge, and some plants have been stolen and introduced into mass cultivation, resulting in some cases (such as coffee and stevia; p. 144) in a multimillion-pound industry. Little, if any, of that money returns to the local people. A report published in 2008 maintained that 74 per cent of the 119 prescription drugs investigated had come to the attention of drug companies owing to their value in traditional herbal healing.

Unfortunately, important herbal knowledge gathered by the Indigenous people of the world is being lost. Colonization eradicated many historical records, and in any case, the knowledge of herbs and traditional medicine has frequently been passed on orally, a habit that is decreasing with each generation owing to urbanization. Such knowledge is not being preserved in scientific research or literature. For example, woad (p. 139) used to be part of a thriving textile and dye industry in Italy, but has been replaced with cheaper synthetic dyes, which are less labour-intensive. Now the knowledge, craft and understanding of the herb are all but lost, as skills have been replaced rather than passed on. We are left merely with the stories of folklore to hint at the importance of specific plants, their healing properties and the processes involved in using them.

The similarities between indigenous herbs are fascinating. Communities that are separated both geographically and culturally have discovered different herbs to treat the same health problems. Garlic (p. 120) and the seeds of cardamom (p. 155) were used to prevent bad breath in ancient Egyptian and traditional Indian medicine respectively. Tea is a drink that is commonly found across the world. People in isolated areas who had not encountered the 'true' tea plant (*Camellia sinensis*; p. 190) developed their own versions of the drink using their native plants, finding in them similar health-giving properties and energy-promoting caffeine. The Guaraní people of South America used stevia to make a traditional green tea; the Inca used lemon verbena (p. 136); and the Bundjalung people of Australia used tea tree (p. 151).

Herbs that were once known only to Indigenous people have spread across the world as food as well as medicine, becoming staple ingredients in many different cuisines. Chilli peppers (p. 147) have become as important in their adopted homes of India, Thailand and Vietnam as they are in their native region of the Americas. The sharing of herbs and the passing on of traditional knowledge are essential for the continuation of Indigenous cultures.

In this chapter I highlight some notable herbs while acknowledging the contribution of the Indigenous people who live with them. In telling these stories, I travel around the world to explore the histories and present-day uses of an array of herbs. Stevia is found in the tropical mountains of Brazil and Paraguay, and aloes (p. 143) grow in the arid landscape of the Arabian Peninsula. The open North American prairies are home to echinacea or coneflower (p. 132), and tea tree is found in swampy areas of southeastern Queensland and New South Wales, Australia. These herbs are brought to wider attention by the Indigenous people who have used them the most.

Coneflowers for colds and coughs

ECHINACEA *ECHINACEA ANGUSTIFOLIA* (ASTERACEAE)
COMMON USE MEDICINE

Echinacea angustifolia grows naturally in North America, across a wide area from the southern United States into central Canada. It does not like competition, so is commonly found in open, dry landscapes with few trees and lots of light. This iconic North American species is particularly popular in ornamental gardening, especially in the steppe and prairie designs that became fashionable in the early 21st century. It has a distinctive orange, cone-shaped centre, which has led to one of its common names. This cone contains hundreds of little flowers that make up an inflorescence. The spiky nature of the central cone is recognized etymologically; the Latin name *Echinacea* comes from the Greek *ekhinos* (sharp points). The flowers are surrounded by purple rays that entice pollinators, such as bumblebees and butterflies. These rays hang down or backwards, rather like the feathers of a shuttlecock.

The North American flora contains more than 17,000 plant species, over 2,800 of which are used by Indigenous people. Echinacea is one of the best known, owing to its popularity and range of applications. It has a long herbal history among the Indigenous people of the United States. First used in about 1600 CE, it was the primary medicine of the peoples of the Great Plains and the Canadian Prairies. They used its roots to treat a number of conditions, such as cuts, wounds, insect stings and snake bites. It was also made into a tea to treat such

diseases as smallpox, measles and mumps (illnesses introduced by the colonists), and ailments such as colds and arthritis. A tea made from the leaves could also be used as a mouthwash to treat toothache. However, recent studies have found that the related species *E. purpurea* is more effective as a mouthwash than *E. angustifolia*, because it has higher levels of the useful chemicals.

Many species in the genus have been used across North America in folk medicine and traditional healing methods. *E. pallida* is used by the Cheyenne and Dakota peoples, who chew the roots for colds, apply them directly to inflamed areas, or take them as an antiviral tincture. The roots of *E. purpurea* are used by the Choctaw to treat coughs and indigestion. *E. angustifolia* is the most diversely used species, however. The roots are ground or smashed into a fine pulp to treat coughs, stomach cramps, bowel pain, snake bites, septic wounds and poisoning, and the juice is used to wash burns and as pain relief. The smoke that arises when the leaves of the herb are burned is used to calm stressed horses.

Despite the differences between the many Indigenous peoples both culturally and geographically, they share similar rules for collecting and gathering echinacea. It is important to collect the herb in the morning, often accompanied by prayers or other sacred words. Many Native Americans depend on the harvesting of plants for survival in the challenging environments in which they live.

Echinacea makes a long-lasting cut flower, sturdy and architectural in shape. Once the flower is over, the cone-shaped seed heads can be used in dried-flower arrangements.

Echinacea does not like competition from other plants,
so in a garden setting it is best given space to itself.

There is no problem with over-collecting, however; the volumes that are taken have always been low – just enough to sustain the community – and they are often not harvested from one single location. This allows time for regeneration after harvest.

The herb began to be used outside the Native American peoples in the late 19th century. A physician from Nebraska named H C F Meyer began selling echinacea root as a blood purifier and a treatment for dizziness. He sent his herbal remedy to John King, a well-known pharmacist based in Cincinnati, who began testing it. King published his findings in the book *King's American Dispensatory* in 1854. He recorded positive results regarding echinacea's treatment of bee stings, blocked noses, leg ulcers and even cholera. Echinacea was later marketed as an anti-infection drug.

By the early 20th century echinacea had gained a reputation as a remedy for colds and flu, and was a key part of the medicine cabinet in many American homes. The advent of antibiotics after World War I caused a brief decline in its popularity, but it was revived in the 1970s owing to its reported safety. That reputation has lasted to the present day. Echinacea is now one of the most popular herbal dietary products in the United States, and is taken in capsules or as a tea.

Despite its widespread availability, there are few studies quantifying echinacea's health benefits. Most focus only on its use as a treatment for coughs and colds, and there has been little investigation into the wider range of recorded uses by Indigenous people. Additionally, there is a notable lack of information regarding maximum dosages and the dangers for people who are pregnant, have other health problems,

or suffer from allergies. Furthermore, the Native Americans primarily use the root, reserving the leaves for specific treatments, while commercially, the whole herb is used.

It is also unclear which species are most commonly used in these herbal remedies. The three species *E. angustifolia*, *E. purpurea* and *E. pallida* are used in variable volumes; sometimes not all are used, and they are often combined with other herbs, which may be giving the benefits for which *Echinacea* is getting the credit. The chemicals present in the herb vary by species. For instance, *E. purpurea* does not contain echinacoside, a chemical that is beneficial to cardiovascular and neurological health. A study carried out in 2005 found that *E. angustifolia* showed significant antiviral qualities, while *E. pallida* had none. These results make two things clear: that the true components of commercial remedies must be examined; and that the health benefits described by Native Americans should be researched further.

Peru's Coca-Cola

LEMON VERBENA *ALOYSIA CITRODORA* (**VERBENACEAE**)
COMMON USES TEA • CULINARY • ANTIDEPRESSANT

This South American herb – a medium-sized shrub that reaches a height of about 2.5m (8ft) – releases a strong lemony scent when its leaves are bruised, and that property has earned it the species name *citrodora*. It is not a reliably hardy plant; winters below freezing will damage it, so it is best grown in sheltered spots or in climates that do not frequently drop below freezing. It is now widely grown in the United States, southern Europe and northern Africa.

The herb is used by South American Indigenous communities. It is thought to have been first employed by the Inca for complaints of the stomach and nervous system. They called it *wari pankara* in their own language, Quechua, and recommended it for muscle spasms, bronchitis and heart problems. It is to lemon verbena's general antioxidant, anti-inflammatory and calming properties that the overall good health of the Inca people can be attributed. In Argentina, Brazil and Uruguay, the herb is used as a mild sedative, to treat insomnia. Its uses in Paraguay tend towards cardiovascular problems, while in Bolivia, an infusion made from the leaves is given to those suffering from 'fear'. The leaves are traditionally steeped as a tea that is used to treat indigestion, colds and flatulence, and is also said to revitalize the body and treat depression.

The plant's lemony fragrance has made it popular in the culinary and cosmetic industries. Its leaves are added to salads for a citrus flavour. The national drink of Peru is Inca Kola, a sweet, fruity soda made with lemon verbena. (Interestingly, Peru is one of the very few countries in the world where Coca-Cola is not the number one soft drink.) Another quintessentially Peruvian drink is the refreshing *emoliente*, which is made with lemon verbena, barley, lime juice, flax seeds and plantain leaves.

Lemon verbena was introduced into Europe by the Spanish. The herb's Spanish common name, *hierba luisa*, commemorates the wife of King Charles IV of Spain, Maria Luisa of Parma, who grew it in the Madrid royal garden in the 18th century. By the end of the 19th century it had become a common tender plant, grown in conservatories and glasshouses across Europe. Its popularity grew among Victorian ladies, who used it as a dried herb. The leaves retain their odour for years and, according to folklore, made one more attractive to potential admirers.

The herb is now commonly used in gardens as an ornamental plant and as a food, for its lemony taste. Its use in aromatherapy has dropped owing to evidence that continued use can increase sensitivity to sunlight, and it has been replaced in some products by the similarly scented lemongrass (p. 205).

Victorian ladies commonly used lemon verbena as a dried herb for the boudoir. The leaves retain their odour for years, and, according to folklore, make one more attractive to potential admirers.

The ancient Egyptians used woad to dye
cloth for wrapping mummies.

Blue tattoo

WOAD *ISATIS TINCTORIA* **(BRASSICACEAE)**
COMMON USES DYE • TEXTILE

Woad is a biennial species in the brassica family, which also contains such notable food crops as broccoli, cabbage, radish and mustard. It has the bright yellow cross-shaped flowers characteristic of the family. It grows in warm, dry, stony landscapes in regions from southeastern Europe to Central Asia, and has been introduced throughout Europe and in similar climates. Despite its tolerance of dry soil, it produces a higher yield when growing in richer, wetter clay soil.

Woad has a well-documented history as both a medicinal and a dye plant. The species name *tinctoria* is a general indicator that a species is used in dyeing, and, indeed, when woad flowers are crushed or the leaves are boiled, they turn a deep blue. The remains of a woman wearing a blue dress dyed with woad were found in Lazio, Italy. This discovery has been dated to the Iron Age, which makes it the earliest evidence of woad being turned into a dye.

Woad is also a common antiseptic and has been used to stop bleeding. Its common association with body-painting and tattooing originates with the Romans, who crossed the Channel to invade Britain and met fierce warriors with blue tattoos. The Romans called them *picti* (painted people). Pliny, who frequently travelled with the Roman armies in the 1st century CE, mentioned that female Britons covered their bodies in the dye for religious celebrations.

The Romans learned the craft of woad dyeing, and as the city of Pompeii recovered from the devastating eruption of Mount Vesuvius in 79 CE, it became the hub of the Roman textile industry. When the empire dissolved, much of the woad industry fell with it, yet embers still burned in southern Italy. The Silk Road trade route between the Western and Eastern worlds expanded, and the dye trade – specifically woad – increased dramatically. The dye continued to be traded until the late 17th century, when the superior dye obtained from indigo (*Indigofera tinctoria*) was imported from tropical regions of the world, replacing woad. The 19th century brought the introduction of synthetic dyes, changing the dyeing industry completely, and woad was largely forgotten in Europe as both a dye and a medicinal plant. Today, it can be found in the cosmetic industry; the seeds are used for oils and moisturizing products, the leaves for hair conditioners, and the root extract for skin protection.

The use of woad in Chinese traditional medicine had no such hiatus. There the leaves and roots are classed as different medicines. The leaves are used to treat flu and mumps, while the root is made into the popular tea *banlangen keli*, which is prescribed for such diseases as hepatitis, SARS (severe acute respiratory syndrome) and scarlet fever.

Korean cure-all

GINSENG *PANAX GINSENG* **(ARALIACEAE)**
COMMON USE ENERGY BOOST

Ginseng's genus name, *Panax,* comes from the Greek *panacea,* indicating that the plant was believed to 'cure all'. The common name is often used to describe several species in the family, among them Siberian ginseng (*Eleutherococcus senticosus*) and American ginseng (*P. quinquefolius*). While these others have been used in herbal remedies, *P. ginseng* – which comes from northern China, Korea and Russia, and is sometimes called Asian or Korean ginseng – is the most used and known. This plant of cool mountain regions has large, hand-shaped leaves surrounding white flowers, which are followed by a cluster of red berries.

Ginseng root has been used for more than 4,500 years. The first-known written account of the plant was by the Chinese calligrapher Shi You in about 48 BCE. It was later included in the *Shennong Bencao Jing* (Shennong's Herbal). The traditional use of ginseng in herbal medicine was as an adaptogen: a natural substance that helps the body to manage stress. Ginseng was also understood to reduce fatigue and improve the circulation of the blood.

The cultivation of ginseng began on a small scale about 4,000 years ago. Plants collected in the wild were transplanted into gardens, a practice that led to a steady decline in natural populations. But by the 11th century CE, production had become sustainable, and wild-growing plants were no longer collected. Ginseng was then grown as a crop and traded with China, where it was in high demand for traditional medicine. There are between four and six years from cultivation to harvest. Fresh ginseng is harvested after four years and can be eaten immediately. White ginseng is harvested after four to six years and must be dried and powdered. Red ginseng is harvested after six years and is steamed, then dried in heat or sunlight.

In the 25-volume Korean herbal *Dongui Bogam* (1610 CE), compiled by the court physician Heo Jun, ginseng was reported as being used in 653 herbal remedies. According to Li Shizhen's 16th-century *Bencao gangmu* (Compendium of Materia Medica), it was used to cure a total of 23 diseases. It has been employed as an antiseptic and as a tonic to increase energy. Ginseng's positive effects have often been attributed to something unique to the genus *Panax*: chemical compounds known as ginsenosides, which vary in concentration according to the species and the part of the plant. Emerging research shows promising results regarding the use of ginseng in treating neurodegenerative diseases, such as Alzheimer's and Parkinson's.

Ginseng is used in a range of products, including food, dietary products, health supplements, medicines and agricultural products. Its popularity has caused the plant to become endangered in the wild. All production is now from nursery-grown material, and nearly all occurs in four countries: the United States, Canada, China and Korea (the largest producer).

The number of uses for ginseng in the present day means
it lives up to its Latin name, *Panax* – panacea, or 'cure all'.

Only the gel of the aloe plant is used; the skin
contains the compound aloin, which is toxic.

Soothing gel

ALOE *ALOE VERA* **(ASPHODELACEAE)**
COMMON USES WOUNDS • COSMETIC

Aloe is a well-known succulent that is a common garden plant in warm, dry climates, and a popular house plant in cooler areas. The family to which it belongs also contains other common house plants, such as *Haworthia*, and such familiar garden plants as the red-hot pokers (*Kniphofia* spp.). Owing to the aloe's long history of cultivation and its journeys along various trade routes, uncertainty surrounds its exact place of origin, but that is likely to be somewhere in the Arabian Peninsula. Aloes have become naturalized in similar climates in the Americas, Australia, India and around the Mediterranean. The word aloe is thought to come from the Arabic *alloeh*, meaning a shiny, bitter substance, and refers to the translucent gel inside the aloe's leaves. *Vera* is from the Latin word meaning 'true', and refers to the confusion that arose in the 17th century between the aloe and the similar agave plant.

The aloe was first recorded on Sumerian clay tablets, then in the ancient Egyptian Ebers Papyrus. It was recommended for infections and skin problems, and as a laxative, and indeed is still used as such. Queen Cleopatra's skincare routine may have involved the aloe plant, which is said to have been the cause of her wonderful skin.

This plant is one of very few non-narcotic herbs to have caused a war. When Alexander the Great took control of Egypt in 332 BCE, he learned of the aloe's healing properties and of its presence on the island of Socotra in the Indian Ocean. Desperate to find methods of healing his soldiers, he sent an army to take the plant and the island. After conquering Socotra, he had large supplies of aloe with which to treat his soldiers' wounds.

Traders carried herbs and spices across the arid regions of the Arabian Peninsula along the famous Silk Road, often using aloes to quench their thirst. The gel that oozes from the leaves has a high nutritional value that was beneficial during the long, arduous journey through the desert. It was also applied to areas of sunburn or as a protective suncream. It was this trade along the Silk Road that brought *Aloe vera* to the countries of the East.

Of course, *A. vera* is best known for that cooling gel, which is effective in the treatment of cuts, burns and other skin complaints. In India, aloe medication is also used to treat stomach ache. In Mexico, the leaves are harvested from naturalized plants and used to treat burns and skin irritation. Many species of aloe have medicinal uses, but *A. vera*'s global use is remarkable. The worldwide market was valued at $1.6 billion in 2018, and that is set to increase. The great part of aloe's economic value comes from its use in skincare, for its reported abilities to heal damaged skin and reduce wrinkles. However, the development of its culinary value as a nutritious drink is helping to expand the range of products that are made from it, and it has recently begun to be added to sugar-free fizzy drinks.

Sweet enough?

STEVIA *STEVIA REBAUDIANA* **(ASTERACEAE)**
COMMON USE SWEETENER

Stevia is a small herb from Brazil and Paraguay, a member of the daisy family. It is native only to the Río Monday valley of the Amambaí mountain region, on the border between the two countries. The plant grows in wet grassland and is an annual species, lasting for only a single year before flowering and dying. Its limited distribution and short life cycle mean that it is under threat, and it is currently classed as endangered.

The first recorded mention of stevia was by the Spanish physician Francisco Hernández in *The Natural History of New Spain* in about 1570, but the herb has been used by the Indigenous Guaraní people for hundreds of years. They call it *ka'a he'e* (sweet herb), and chew the leaves as a treat. Stevia is also used as a flavouring, as well as a sweetener in the Indigenous herbal green tea *yerba mate* (itself made from a plant in the holly genus, *Ilex paraguariensis*), which is high in caffeine and has a pungent flavour with a bitter aftertaste.

Stevia has religious significance for the Guaraní people. During the ceremony in which a boy becomes an adult, he will wash his hands and arms in a cold infusion of the herb to grant him success with his crops. The Guaraní also use it as a medicinal plant to treat parasites, aid digestion, and act as an antiseptic and astringent.

The biological properties of stevia were first described in 1887 by Dr Moisés Santiago Bertoni, director of an agricultural college in Asunción, but it was not until 1900 that the Paraguayan chemist Ovidio Rebaudi isolated the active ingredients of the herb in order to investigate what made it sweet. His efforts were immortalized in the species name, *rebaudiana*. He found that the leaves and stem cells of the plant contain a chemical that was later given the name 'stevioside'. It is far sweeter than sugar – between 200 and 300 times – and was first used as a sugar substitute in the 1970s.

Stevia has no calories and does not break down in the human body, meaning that it is safe for those, such as diabetics, who monitor their blood sugar. The World Health Organization currently recognizes stevia as a bacterial agent and blood-pressure regulator, and no negative health side effects have been reported. In fact, it is regarded as a healthier alternative to synthetic sweeteners, such as sucralose and aspartame.

When companies began to consider sugar alternatives owing to the 'sugar tax' (or 'sweetened beverage tax') imposed in many countries in the early 21st century, several turned to stevia. In 2012 stevia-based Coca-Cola and Pepsi became available. In 2015 the Guaraní people submitted a lawsuit against some of the world's major soft-drinks companies, on the premise that the herb had been taken from the Indigenous people and used to build an industry worth some $500 million a year. None of that money has come back to the local people.

The genus *Stevia* is named after Pedro Jaime Esteve,
a botanist at the University of Valencia in the 16th century.

Chillies can add heat to all kinds of recipe, and are often used as
condiments, as in the famous Thai hot sauce sriracha.

Mexican chillies

CHILLI PEPPER *CAPSICUM ANNUUM* **(SOLANACEAE)**
COMMON USE CULINARY

Capsicum annuum is the best-known of the five species of chilli pepper that are commonly cultivated; the others are *C. chinense*, *C. frutescens*, *C. baccatum* and *C. pubescens*. Its native range was originally coastal and southern Mexico, but it has now spread to Guatemala, where it grows naturally in wet areas. Owing to its long history of domestication, there is great variation in the species. Sweet (bell) peppers, paprika, cayenne and Mexican chillies are all variations of *C. annuum*, yet the plant typically looks the same: short, herbaceous, with small white flowers. It shares a family with potato, tomato and aubergine (eggplant).

The genus name, *Capsicum*, comes from the Latin *capsa* (box), and refers to the way the fruit encloses the seeds. The common name, chilli, is lifted directly from the Aztec language Nahuatl.

Chilli peppers of various *Capsicum* species have been cultivated in the Americas since about 4000 BCE, and archaeological finds from caves near the Tehuacán Valley in Mexico strongly suggest that *C. annuum* was harvested as early as 2,000 years before that. This was a period in which early hunter-gatherers turned to crops and farming. They stopped harvesting wild plants, instead planting them as crops and cultivating them. Such plants as avocado, corn and beans were grown to feed small farming families. They travelled with these crops to hunt deer and to collect wild plants, such as agave and cactus fruit, often sheltering in caves when they were too far from their villages. It was in these caves that chilli peppers were used, perhaps to flavour food or to reduce hunger – an attribute that they have been shown to have.

In the late 15th century the Ottoman empire took control of the pepper trade from the Malabar Coast in India to Europe, and Christopher Columbus sought a new trade route to Asia, principally to renew the stream of 'black gold' (black pepper; p. 210) for the Europeans. However, instead of landing in India, he reached the Caribbean islands, where the Indigenous people had been using chilli peppers for centuries. Columbus named the *Capsicum* fruit *pimiento*, after black pepper (*pimento*), and that is why they are to this day called peppers instead of the local name, *ají* (which is from the Taíno language).

Columbus introduced chilli peppers to Spain, where they were at first not appreciated for their strong, spicy flavour. After a while, however, people became accustomed to them and began to use them in cooking and to investigate their medicinal value. Chillies were spread across Europe and Asia by Spanish and Portuguese trading ships, along with the knowledge of how to cook with them, and of their medicinal qualities. They are used in various cultures as a traditional medicine, being understood to relieve the pain of headaches and arthritis and to clear coughs and blocked noses. In the Unani medicine of southern and central Asia, they are used to improve digestion and to prevent colds and sore throats. The antioxidant, antiviral

and anti-inflammatory properties of chillies are good for general health and well-being.

The bright colours and sometimes painful spiciness of chillies are an evolutionary attempt to discourage mammals (including humans) from eating them. In the plant kingdom, red is a colour that attracts birds to flowers for pollination or to fruit for dispersal, and, since birds do not have the same pain receptors as mammals, they can eat chillies without experiencing the burning taste that humans do. It is the chemical capsaicin that causes this notorious sensation, and if too much is eaten at once, it can cause health problems. Capsaicin is a key component of pepper spray, which was first developed by the FBI in 1987 and is used by law-enforcement officers to incapacitate dangerous suspects or as a method of riot control.

A chilli's heat is measured using the Scoville scale, which is based primarily on the concentration of capsaicin. Hot chillies have high Scoville numbers. For example, sweet peppers are rated as zero Scovilles, while jalapeños measure between 2,500 and 10,000 Scovilles. The Carolina reaper – named the hottest chilli pepper in the world in 2017 – is recorded at 1,641,183 Scovilles. It is a hybrid between two selections of *C. chinense* and was bred specifically to become the hottest pepper in the world.

The researchers David Julius and Ardem Patapoutian were awarded the Nobel Prize in Physiology or Medicine in 2021 for their investigation into the pain receptors triggered by capsaicin. It was through chilli peppers that they discovered the molecules that are responsible for humans' ability to sense temperature and touch. If you want to avoid triggering your pain receptors with the intense heat of a chilli, there is a way. Many of us have learned to remove the seeds of chillies to make them taste less hot, but in fact capsaicin is contained in the inner white pith or ribs of the peppers. The seeds contain little or no heat, so to make a dish milder, it is the ribs that should be removed. Another tip is to temper the heat of chilli with dairy products, which contain casein, a milk protein that neutralizes capsaicin.

Chilli peppers are an important ingredient in authentic Indian, Vietnamese, Thai, southern Asian and East African dishes. They can be used both fresh and dried, and in oils, marinades and dry rubs. They are often added to sauces or broths to increase the heat. Many Mexican dishes feature different chillies. In Chile, a fresh salsa called *pebre* is made using traditional *ají* peppers, olive oil, fresh coriander/cilantro (p. 17) and garlic (p. 120). The level of spice and consistency varies across the country.

The medical value of chilli peppers has been analysed and evaluated in recent times. They are shown to have preventative qualities against diabetes and cancer, while also improving overall heart health and blood pressure.

Despite a widespread belief to the contrary, the seeds of the chilli contain little or no heat; it is rather the white, fleshy pith that provides the spiciness.

Tea trees are commonly grown as boundary hedges in Australia to prevent soil erosion.

Aboriginal tea

TEA TREE *MELALEUCA ALTERNIFOLIA* (MYRTACEAE)
COMMON USES ESSENTIAL OIL • COSMETIC

The tea tree is native to parts of Australia – chiefly southeastern Queensland and the northeastern coast of New South Wales – where it grows in swampy areas and along the banks of rivers. A relative of eucalyptus and myrtle, it forms a small tree with thin, needle-like leaves and plumes of white candyfloss flowers.

It is not clear when Bundjalung Aboriginals – an Indigenous people who live in southeastern Queensland – began using the tea tree. There are several reasons for this uncertainty. First, the common name 'tea tree' is applied to all species of *Melaleuca* and various other genera. Additionally, several other species were used for similar traditional healing practices, and very little Indigenous knowledge has been preserved in written form.

In the 1770s a record of the Bundjalung people was created by the botanist Joseph Banks, who had been a member of the British explorer Captain James Cook's crew on his famous voyage of 1668–71. Banks observed the Bundjalung people's understanding of medicinal herbs and how they used tea tree leaves to cure certain ailments. The exploring sailors consumed the leaves with water as a substitute for tea, a habit that is probably the origin of the common name 'tea tree'. The Bundjalung people use the plant for a range of medicinal applications, however, including treating colds, sore throats, headaches, insect bites and wounds, and for delousing.

In 1922 the Australian chemist Arthur de Ramon Penfold extracted oil from *M. alternifolia*. He maintained that the plant's essential oils had a range of antimicrobial and antifungal properties that could treat various infections, and his findings generated international interest. However, it was decreed that all available essential oils be sent to treat war wounds during World War II, so tea tree remained largely in Australia until the 1990s, when new studies reignited interest once again.

Tea tree oil is used today in creams and washes to treat infections, such as athlete's foot, and to reduce acne and relieve irritated skin; the essential oil can also be added to a bath, ideally with a neutral carrier oil. It is not administered orally, owing to its toxicity. It has also been used since the turn of the millennium as an antimicrobial cleaner in the poultry industry. Again, its toxicity means that it is used for animals only with extreme care.

European ginger

ASARUM *ASARUM EUROPAEUM* (**ARISTOLOCHIACEAE**)
COMMON USE TRADITIONAL MEDICINE

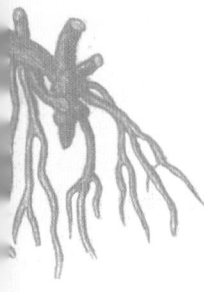

Asarum is an unusual and attractive woodland genus belonging to the birthwort family. The kidney-shaped evergreen leaves have fine white hairs, and the flowers are cherry-red. As with many woodland plants, the seeds are dispersed by ants. *A. europaeum* is widespread throughout Europe, from Italy all the way to northern Russia. When the roots are broken or unearthed, they emit a strong gingery, peppery odour, a property that has earned asarum the common names 'European ginger' and 'ginger root', even though it is not closely related to the true gingers (*Zingiber* spp.; p. 194).

During the Middle Ages, asarum was recorded in Polish folklore for use during the most important day of the Catholic calendar, Corpus Christi, when it was taken (along with other plants, such as *Alchemilla*, *Achillea* and roses; pp. 53, 100, 179) into the home to form a tribute to God. Since *asa* means altar or sanctuary, this practice is probably the origin of the Latin name *Asarum*. Hildegard of Bingen recommended adding the herb to a bath to relieve skin irritation, and in the 16th century another German physician, Christoph Wirsung, recorded asarum as a treatment to purge the black bile generated by quartan fever, a type of malaria.

Asarum has long been used in Unani medicine. The plant is not native to Central Asia, but Arabic physicians and scholars learned of its medicinal properties from their Greek counterparts, and used it in the treatment of a wide range of illnesses. The roots are broken down into a decoction that is used to treat epilepsy, sciatica, and stomach and liver disorders. The famous 19th-century Unani physician Hakim Mohammad Azam Khan Chishti maintained that the best roots were those of medium thickness, and that thin roots could cause tingling and inflammation in the mouth. The leaves, meanwhile, have been recorded as an antidote for snake bites and scorpion stings. Asarum is often mixed with other herbs; it is a key ingredient in *dawā al kibrit*, for example, a herbal remedy for digestive disorders. According to Unani folklore, asarum also builds nerves, muscles and general strength.

The herb is not entirely safe, however. High doses can cause fever, nausea and internal bleeding, and can even be fatal. Asarum contains the chemicals asarone and aristolochic acid, which are carcinogenic in high doses. In fact, Unani medicine has developed remedies for asarum overdoses. Almond (*Prunus amygdalus*) and grape (*Vitis vinifera*) are added to asarum to minimize these side effects, while myrrh (*Commiphora* spp.) is an antidote that operates by returning the body to a safe temperature. The plant's toxicity has caused it to fall out of favour in recent years. However, it remains a common ornamental plant for the garden, where it grows well in shaded areas.

Asarum was recorded as a perfume in ancient Greece, and was added
to medicinal drinks for its peppery, gingery taste.

The common name 'cardamom' comes from the Greek *kardamon*, meaning 'cress'.

Indian breath freshener

CARDAMOM *ELETTARIA CARDAMOMUM* (ZINGIBERACEAE)
COMMON USES CANCER • CULINARY • BREATH FRESHENER

Cardamom is native to tropical southern India, where it grows in dry areas. *Elettaria* comes from *elattari*, the plant's name in the language of the Indigenous Tamil people, who live in the southern Indian states, including Kerala and Karnataka, where it grows. These are also the main areas of cardamom cultivation in India. Cardamom belongs to the ginger family, but, unlike ginger (p. 194), it is not the tuber that is eaten but the fruit.

The herb is mentioned in the ancient Egyptian Ebers Papyrus as having seeds that were chewed to clean the teeth and freshen the breath. The plant is also said to have grown in the Hanging Gardens of Babylon in about 700 BCE. Cardamom was used by the ancient Greeks and Romans, and appears in Ayurvedic medicine. Its applications covered a range of conditions, from kidney disorders and asthma to teeth and gum infections. It was also used as a diuretic. Greek physicians, from Hippocrates in the 4th century BCE to Dioscorides in the 1st century CE, regarded the herb highly as food, medicine and perfume.

In Kerala and Tamil Nadu in southern India, cardamom pods are boiled with black tea (*chai ki patti*) and water to make *elakkai* tea. The fresh, pungent aroma of cardamom (which has light undertones of citrus, pepper and camphor) passes into the tea, which is said to treat tiredness and depression. Boiled cardamom pods are also used to make the fermented alcoholic drinks *asava*

and *arishta*. Cardamom is mixed with honey and milk and drunk to improve the eyesight in traditional Ayurvedic medicine, and is regarded in Indian traditional medicine as an excellent herb for digestion. In traditional Tibetan medicine, cardamom is taken with cinnamon and long pepper (*Piper longum*) to treat diseases of the liver, kidney and heart.

Cardamom, the 'queen of spices', is the third most expensive spice in the world, after saffron and vanilla. It adds flavour to curries and other Indian dishes in the form of the spice blend garam masala, and is included in Swedish cardamom buns (*kardemummabullar*) and the crunchy Dutch biscuits *speculaas*. In the Middle East, cardamom is taken with *gahwa* coffee to lessen the effect of the caffeine.

The production of cardamom in tropical climates has accelerated in recent years, and Indonesia has overtaken India as the top producer. Cardamom plants are cultivated in the temperature range 10–35°C (50–95°F), and the dry fruit is harvested every three years. The top consumer countries are Scandinavia, Iceland, Russia, the United States and Japan. Cardamom's medicinal properties have been scientifically proven in recent years, and additional ones are being investigated, most notably a possible role in treating breast cancer. The herb contains various antioxidant, antibacterial, anti-inflammatory and immune-boosting compounds that have been shown to destroy cancerous cells in patients.

Rose; see p. 179.

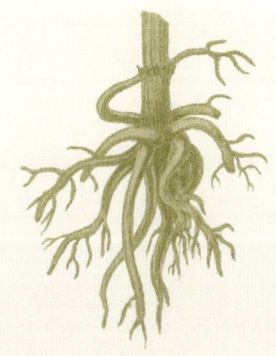

MONASTIC HERBS

Monastic Herbs

As the Roman empire fell in the West in the 5th century CE, European medicine was adopted by the Catholic Church. Monks learned from the works of the ancient Greek and Roman scholars, allowing them to develop and preserve the knowledge of the past. Following the instruction of St Benedict, monks began to curate gardens, cultivating various herbs for medicinal and ornamental use. Benedict argued that the monasteries should be entirely self-sufficient, providing themselves with crops for food, access to water, and craft skills. This independence also involved the growing of medicinal plants to establish a sustainable healthcare system within each monastery. The Latin species name *officinalis*, used for many medicinal plants, denotes this particular use. It translates literally as 'belonging to an *officina*' – the monastery's medicinal store.

In 787 CE Emperor Charlemagne visited the Benedictine monks at the abbey of Monte Cassino, Rome. He admired the herb gardens and medicinal plants there so much that he ordered all monasteries in his empire to follow suit. The number of monastic gardens consequently grew from a few hundred to several thousand. They spread throughout Europe and far north to Greenland, Iceland and Scandinavia. The herbs grown in monastic gardens varied according to the climate and situation of each, and monastery gardens in the north needed hardy herbs to survive the cold winters. As we have seen, angelica (p. 66) was a particularly important crop in northern European countries. French and German monastic gardens, on the other hand, could grow Mediterranean herbs, such as chamomile, sweet cicely and lavender (pp. 163, 176, 180).

The monks became keen gardeners. As well as promoting self-sufficiency, they planted species of ornamental value as a tribute to God. These gardens were enjoyed as peaceful places that helped the monks in their contemplation and relaxation. As the monks travelled to other parts of the world, they gathered knowledge of different cultures, ways of life and materials, searched out new herb species and shared the ones they had. A network of knowledge and exchange was established globally, and some monastic plants have spread outside their native areas for this reason. From Arab scholars, European monks learned to add their health-benefiting herbs (sweet cicely, for example) to alcoholic drinks, such as the famous green liqueur known as chartreuse.

Herbal knowledge was also spreading around mainland Europe. When settlers moved to England from Germany, taking their plants with them, they established new herbs, such as lavender, in the south of the country. In exchange, they learned about the sacred plants and healing herbs used by the Anglo-Saxons and native Druids.

Despite their lesser status at the time, women in the Middle Ages were essential to medicinal practice through their roles as midwives and nurses. One of the most notable monastic gardeners was the 12th-century German Benedictine abbess Hildegard of Bingen. A nun from the age of 15, she compiled some of the most influential herbal books of the Middle Ages. She recorded a range of culinary and medicinal uses for monastic herbs, including valerian and southernwood (pp. 168, 184).

Some monastic herbs are still very common today. Herbal teas made from chamomile and southernwood are easy to find in health-food shops and supermarkets, and offer a range of health benefits. The aromatic and culinary properties of monastic herbs are still being used by monks throughout Europe. The sale of therapeutic products (such as those made with hand-harvested lavender) and alcoholic herbal drinks supports the life of these monasteries. Appreciation of the beauty of plants was an important aspect of monastic gardens, and many monastic herbs have become ornamental species, among them catmint (p. 175) and lavender. In this chapter I explore some of the best-known monastic herbs, their history and their uses.

Biblical-strength cleaner

HYSSOP *HYSSOPUS OFFICINALIS* **(LAMIACEAE)**
COMMON USE DRINK

Hyssop is an aromatic shrub that shares a family with rosemary, oregano and lavender (pp. 30, 37, 180). It also has a similar distribution to these herbs, being found mainly in an area stretching from the eastern Mediterranean to Central Asia. It has long, thin leaves and fragrant white flowers and, like its close relatives, contains essential oils that are used in medication, herbal remedies and cooking.

The use of hyssop is estimated to have begun around 2,500 years ago. The herb symbolized purity and forgiveness, and the first known record was by Jewish priests, who used it to clean their places of worship. Its cleaning properties are referenced in the Bible – 'Purge me with hyssop and I shall be clean' (Psalms 51:7) – although it is possible that it was misidentified, given its similarities to other genera in the family Lamiaceae; one of the common names for another relative, *Origanum syriacum*, is 'Bible hyssop'.

The ancient Greeks used hyssop in the form of a tea to treat colds, coughs and chest problems. Dioscorides recommended smelling the herb to clear nasal and chest congestion. Much later, in the 12th century, Hildegard of Bingen suggested that it could help to clear the lungs and, if drunk with wine, alleviate depression.

Hyssop has been recorded in numerous monastic gardens throughout Europe. Its distinctive taste and clean, minty scent made it popular as a culinary herb and preservative, and it is one of the key flavours in the famous liqueurs Bénédictine and chartreuse. Bénédictine is said to have been made first by the Benedictine monks of Fécamp Abbey on the Normandy coast. During the Middle Ages, hyssop was recorded in Norwegian and Icelandic monastic gardens. It was also one of the earliest Mediterranean herbs to reach Russia, where it was recorded as early as the 1350s in Slavic monastic gardens.

Like Hildegard, Nicholas Culpeper endorsed hyssop as a herbal remedy for colds and chest pains, explaining that 'it expelleth tough phlegm and is effectual for all griefs of the chest and lungs.' It was also used as a strewing herb throughout the Middle Ages. Introduced by settlers to North America, it was never very popular as a medicine. However, in the 19th century physicians there prescribed it for pain relief and as a treatment for asthma, chest infections and coughing.

Today, hyssop essential oils are used in perfume, and the herb's primary medical use is to relieve gastrointestinal complaints. In cooking, it is most commonly used to flavour meat and salads, and it continues to feature in liqueurs. In addition, it is often used by honey producers as a bee plant, passing its aromatic minty flavour into the honey for a unique taste.

Micro-organisms that produce the antibiotic penicillin have been found on hyssop,
resulting in the belief that the plant can cure chest infections and heal wounds.

The leaves of chamomile are best harvested in the summer,
when they are dry, and after the flowers have begun to droop.

Steeped in history

CHAMOMILE *CHAMAEMELUM NOBILE* (ASTERACEAE)
COMMON USES TEA • SEDATIVE

Chamomile is a perennial ground-cover species that grows in dry areas. It is a native of Morocco and Algeria that – having been introduced hundreds of years ago – has long been considered native to England and Ireland, too. Owing to its distribution, it has two distinct forms. The northern African forms are adapted to drought, while the English and Irish forms can grow in wetter soils and cooler climates. The latter populations are now decreasing through lack of grazing, the drainage of wetlands and the illegal collection of wild plants. Chamomile has also been introduced into the rest of Europe, the east of the United States, Pakistan, Colombia and Australia.

A member of the daisy family, chamomile produces small fern-like leaves and, during the summer months, lots of starry flowers with glowing yellow centres and bright white petals. It is a herbaceous perennial, meaning it comes back every year from underground growth.

Chamomile shares its name with German or Hungarian chamomile (*Matricaria chamomilla*), which has a wider eastern distribution. The two plants are similar in appearance – although German chamomile is an annual and about twice the size of its relative – and they produce a comparable light blue oil. The common name of both is derived from the Greek *chamaimelon* (earth apple), a reference to the plant's low stature and slight apple scent. Since the African species was understood to

have superior medicinal, therapeutic and overall health-improving qualities, it was given the species name *nobile* (noble).

The use of chamomile can be traced back to the ancient Egyptians. We know from the Ebers Papyrus that it was used to treat fever caused by malaria and sunstroke. The latter use may have come about from the resemblance of the flowers to little suns, and indeed chamomile became the sacred plant of the sun god Ra. It was given as an offering to the gods and was part of the embalming process, as well as being used to improve sleep.

Chamomile was recorded in the writings of the ancient Greek and Roman physicians Hippocrates, Dioscorides and Galen. The flowers were crushed and used to revitalize dry skin and to heal wounds or cuts, and it was also used widely as a mild sedative, normally in the form of a drink. The famous 1st-century surgeon and gynaecologist Asclepiades of Bithynia advocated for the soothing use of chamomile, believing the flowers, leaves and roots of the plant had a warming, relaxing effect on his patients. At about the same period, Pliny recommended chamomile to treat headaches and kidney, liver and bladder problems.

Every medieval household contained several essential herbs for the treatment of sickness or wounds. Chamomile was one such herb, taken as a tea principally for digestive problems and also to treat fatigue, manage pain, and as a diuretic. It was also used as a strewing herb, its apple scent repelling insects and improving the overall odour of the home. It was believed to help the dead rest, and was planted on graves to prevent the return of the spirit after death. This last practice may have been passed on from the ancient Egyptian use of it in embalming.

Chamomile was one of the nine treasured herbs of the Anglo-Saxons, said to have been sent by the Nordic god Odin. The plant was recorded in the *Lacnunga*, a 10th-century collection of medical texts, for its calming effect on mind and body. It was equally valued for monastic gardens. Monks added the flowers to infusions to ease digestion and diminish flatulence, and drank chamomile tea just before bed as a sedative, to help them relax and improve their sleep.

In the 16th century John Gerard wrote that chamomile 'is exceedingly good against all manner of ache and paine'. It was often added to wine, and Gerard advocated for its use against colds and flatulence. In the following century Nicholas Culpeper recommended the herb for fever, aches and pains, and to promote menstruation. The British took the plant in the 20th century to North America, where it was used to treat gangrene, typhus and birth-related difficulties, from premature labour to foetal kicking. In the early 1930s Maud Grieve recommended chamomile to treat nightmares, among many other uses.

Chamomile tea is popular for its relaxing, soothing qualities. Globally, some one million cups of it are drunk every day. A Japanese study found that just one serving per day decreased the heart rate and lessened feelings of sadness and depression in young men. Studies have also proved that chamomile extract has significant wound-healing abilities. The herb is famously used for its essential oils, including in aromatherapy. In the cosmetic industry, chamomile essential oils are commonly used in skin washes and gels, as well as hair-lightening products.

Over its history of recorded use, chamomile has been used to treat scores of complaints, and indeed Varro Tyler hailed it as 'capable of anything'. Research carried out in 2021 proved that chamomile has a range of health benefits, including the easing of inflammation, gastrointestinal problems and anxiety. Further research is needed to investigate the validity of the large range of claims made for chamomile's efficacy in traditional medicine, as well as to evaluate more uses for the herb, but in the meantime its popularity as a calming tea is sure to increase.

Dry or fresh chamomile flowers can be harvested
by combing the fingers through the plant.

The name 'motherwort' originates from this herb's use during childbirth, when it has traditionally been given to ease the birth by relaxing the uterus.

Calming and soothing

MOTHERWORT *LEONURUS CARDIACA* (LAMIACEAE)
COMMON USES HEART MEDICATION • ANTIDEPRESSANT

Motherwort is now considered native to an area stretching from Europe to Central Asia, although it probably originated in the latter. A colonizer of freshly disturbed areas, it is consequently common along roads, on waste ground and in hedgerows, and although it prefers shade, it is tolerant of other conditions. The light green leaves, with their three narrowly pointed tips, resemble dinosaur footprints, while the fluffy pink flowers that encircle its square stems are appealing to wildlife, particularly bees.

The plant has a long history of use in the treatment of depression and anxiety. The ancient Greeks used it to reduce heart palpitations, and gave it to pregnant women to calm their nerves and stimulate labour, hence its common name. Motherwort has a rich history in ancient China, where it is symbolic of longevity; according to legend, a child was banished from his village and found sanctuary in a small valley, where he survived, eating only motherwort, for 300 years.

By the Middle Ages, the herb was being cultivated outside its natural range. It is recorded in herbals from Hungary, Romania, Russia and Bulgaria, a sign of its establishment as a medicinal plant. Its primary uses were as a treatment for heart ailments and as a sedative for nervous conditions, and its supposed uplifting effect meant that it was often planted in monastic gardens. It is long-lived, and relic populations have been found in abandoned monasteries and churchyards across Europe, its presence in the latter being a sign of its use to repel demons, evil spirits and black magic through its strong scent. Maintaining the beliefs of the Middle Ages, John Gerard recommended motherwort in the 16th century for heart problems, and later Nicholas Culpeper wrote, 'there is no better herb to take melancholy vapours from the heart.'

The species name *cardiaca* comes from this association with the heart. Indeed, the extensive recorded history of motherwort's usefulness in heart medication is now supported by studies showing that it can reduce blood pressure and alleviate heart-related problems. Additionally, a study in 2011 yielded positive results regarding its use for anxiety; although some individuals reported minor side effects, 32 per cent showed a significant improvement and 48 per cent a moderate improvement in feelings of anxiety.

Motherwort has a complex chemical composition that gives it anti-inflammatory, antimicrobial and cardioprotective properties. It is available in various forms from health-food shops, and recommended for treating a range of problems, from diarrhoea to stomach irritation. The fresh leaves can be infused into a tea that is taken mixed with honey or lemon. However, the presence of the chemical leonurine, which can induce uterine contractions, means that (despite the Greeks' early use) it should not be taken when pregnant.

Nature's chill pill

VALERIAN *VALERIANA OFFICINALIS* (CAPRIFOLIACEAE)
COMMON USES SEDATIVE • ANTI-ANXIETY

Valerian grows to 1.5m (5ft) tall and is a perennial commonly found on roadsides, in woodlands and by rivers in an area reaching from Europe to the northwestern tip of Iran. It produces a cluster of small pink or white scented flowers that are particularly attractive to hoverflies, butterflies and bees.

The genus and common names may derive from the name of the Roman emperor Publius Licinius Valerianus, or perhaps from the Roman province of Valeria (in present-day Hungary and Croatia). The ancient Greeks and Roman scholars recorded a range of uses: Dioscorides as a diuretic, Pliny for pain relief, and Galen as a decongestant. Such varied use is hinted at in the old German name for valerian, *baldrain*, thought to be a reference to Baldur, a hero-god of Norse mythology, who was immune to everything (apart from mistletoe). This gave the plant the reputation of an all-healing herb that could cure any illness.

Valerian is first recorded in monastic gardens in the early 9th century, although it is likely to have been used there earlier than that. Hildegard of Bingen was particularly fond of it, and grew it in her monastery at Disibodenberg, southwestern Germany, in the 12th century. She recommended valerian as a sedative and sleep aid, and added it to an elixir to treat a sickness known as *Vich* (possibly a form of cancer). The elixir was, essentially, an early version of chemotherapy used to treat early-stage cancer. Much later, in the 19th century,

the Catholic priest Sebastian Kneipp used valerian as an anti-anxiety medication and to suppress nervousness. In the early 20th century German women carried small bottles of valerian extract to calm any nerves that might arise if they were approached by a potential suitor.

According to 13th-century folklore, the Pied Piper of Hamelin was asked to lure rats from the German town by playing his flute. It is said that he also carried valerian root with him, using its hypnotic qualities, along with the sound of his music, to entrance the rats. Interestingly, valerian root does contain similar chemicals to catnip, which can attract cats and other mammals.

Valerian contains active chemicals called valepotriates, which are in higher concentrations in the fresh plant, and can be carcinogenic. These chemicals are broken down and released during the drying process, making the herb safe to consume. After it is dried, valerian can be taken as a tea to calm and relax, thus reducing pain. The herb is most widely recognized today as a sedative to relieve stress and insomnia. While it does not work instantly to alleviate these problems, it does have relaxing qualities that – over a period of days or weeks – can help. The plant can also be found in gardens as a tall, attractive ornamental that is beloved of pollinating insects.

Essential oils from the valerian plant have historically
been used to calm stressed animals.

In Ukraine, there is a tradition of rinsing the hair with
an infusion of lovage leaves in order to attract a partner.

Love potions

LOVAGE *LEVISTICUM OFFICINALE* (**APIACEAE**)
COMMON USE CULINARY

This large herb – a relative of carrots, fennel (p. 18) and sweet cicely (p. 176) – can grow to 2.5m (8ft) tall. It produces a large rosette of leaves and, in the summer months, bright yellow flowers as an umbel: a group of small umbrella-shaped flowers. It is native to southern Iran and Afghanistan, and it has become naturalized in Europe, North America and eastern Asia.

The ancient Greeks chewed lovage seeds to aid digestion and decrease flatulence. They held the herb in high regard, and gave it as an offering to the goddess Aphrodite. The Romans used the herb similarly, and later introduced it to England, where it was cultivated in monastic gardens. Lovage was grown in the gardens of Norton Priory in Cheshire, England: the most excavated monastic site in Europe. The monks there used it to treat a disease called 'priory' – now known as Korsakoff syndrome – a non-progressive form of dementia commonly caused by alcohol abuse. A tincture of lovage, elecampane (p. 112) and garlic (p. 120) was also made in monasteries, to treat tuberculosis.

The herb was originally called 'love parsley' – *loveache* in Middle English and *luvasche* in Anglo-Norman – because of the similarity of its leaves to those of parsley (p. 21). In the Middle Ages, lovage was associated with love potions and spells. Women added its leaves to baths to induce lust; sprigs were added to clothing to increase the wearer's attractiveness; and it was understood to be an aphrodisiac. It was also connected to superstition, and as such was fed to cattle on St John's Day in midsummer to protect them from evil. More prosaically, the leaves were placed in shoes to act as an antiseptic and keep them smelling fresh.

In the late Middle Ages, lovage fell out of favour and was replaced by new spices and herbs from other countries. However, Nicholas Culpeper in the 17th century still championed its medicinal properties. He believed that the seeds were the most powerful part of the plant, and that an infusion made from them, 'Being dropped into the eyes[,] taketh away their redness or dimness … It is highly recommended to drink the decoction of the herb.' He also recommended the leaves to treat boils: 'bruised and fried with a little hog's lard and laid hot to any blotch or boil [to] quickly break it'.

Today, lovage is used to treat migraines, gout, and bladder and kidney problems. It can also be found seasonally in farmers' markets and greengrocers in Britain and southeastern Europe. The entire plant is edible: the dried roots and leaves can be added to stews and soups; the stalks can be candied; and the seeds are spread on bread or used in alcoholic drinks, such as the old-fashioned lovage cordial, which is used to cure hangovers. Lovage tastes like celery, but sweeter and with a stronger flavour. It has tones of anise and parsley, and the strongest flavour is in the leaves.

Snake heads for snake bites

BIRTHWORT *ARISTOLOCHIA CLEMATITIS* (ARISTOLOCHIACEAE)
COMMON USES SNAKE BITES • GOUT • WOUNDS

This fine-looking plant, with green heart-shaped leaves on a vine-like stem, produces clusters of yellow pipe-shaped flowers underneath the leaves. The flowers are partly reflexed, with the appearance of a snake's open mouth just before it strikes. The plant can grow either erect – to about 1m (3ft) high – or prostrate, along the ground. A bulb-like underground structure helps it to survive in dry conditions, allowing it to remain dormant for long periods, until conditions are more favourable. Birthwort can now be found throughout Europe, typically in stony areas with some drainage.

The medicinal history of birthwort begins with the ancient Egyptians. Theophrastus wrote in the 3rd century BCE: 'It is an excellent treatment of head wounds, good likewise for other wounds (or ulcers) and against snakebites and to induce sleep, and for the uterus made into a pessary.' It was recorded as part of a herbal tincture, with galingale (*Cyperus longus*), wormwood (p. 74), cumin (p. 201), salt and honey, that was used to 'clean' the female reproductive tract and thus functioned as a traditional method of birth control and abortion. As such, it was much used by prostitutes. Conversely, the herb was also an aid in childbirth. This is recognized in both the Latin name *Aristolochia*, 'best birth', and the common name.

In the early Middle Ages birthwort started to grow in popularity as a childbirth aid and as a gout medication. The plant spread around Europe and began to be grown in monastic gardens. Hildegard of Bingen taught that ingesting the herb would help to 'open the female inner organs' and dissolve hardened menstrual blood. Birthwort was introduced into Britain and began to be grown as a medicinal plant in monasteries there. The plant is seldom found in old monastic gardens, although it is still present among the ruins of the 12th-century abbey of Godstow in Oxford. Rather, it has escaped into the wild in a few places, although it has never become invasive.

Aristolochia clematitis is toxic and is used today only in rural communities in eastern Europe, particularly Romania. It contains a chemical compound called aristolochic acid, which, in the correct dosage, can increase the activity of white blood cells and thus speed up the healing of wounds. In the 1950s, however, a string of small communities along the Danube River from Croatia to Bulgaria developed severe health problems. It was discovered that birthwort – which commonly grew in the wheatfields there – had been caught in harvesting machines and made into bread. Repeated consumption caused a build-up of aristolochic acid and resulted in various ailments, from tumours to kidney failure. The conditions caused by this overconsumption became known as Balkan endemic nephropathy. About 100 cases were recorded in Belgium as recently as 1992, when a weight-loss supplement was found to contain trace elements of aristolochic acid.

The newly hatched larvae of the pipevine swallowtail butterfly of North and Central America
eat the leaves of some *Aristolochia* species, which makes them poisonous to predators.

Catmint is a common ornamental plant that is found
in many gardens and city plantings.

A cat's best friend

CATMINT *NEPETA CATARIA* **(LAMIACEAE)**
COMMON USES TEA • CAT TOYS

Nepeta cataria is a tough perennial species in the mint family. It grows in a wide range of habitats over a large area, being considered native from southern Europe to central Honshu in Japan. It is now widespread in North America, and can also be found in South America and New Zealand. This common ornamental garden plant for dry climates is popular for its strongly scented silver foliage and attractive flowers.

Anyone who has a cat will surely be familiar with catmint or catnip. An old gardening rhyme goes: 'If you set [plant] it, the cat will eat it, if you sow it [as a seed], the cats don't know it.' Certainly, if it is planted as a grown plant, the leaves release a fragrance that attracts cats to eat and roll around in its leaves. Catmint contains an essential oil called nepetalactone, which elicits this euphoria. It is still unclear how, but cats do seem to respond to it as though it were a pheromone. Recent studies have suggested that catmint has a stronger effect on adult cats, and that it causes a dopamine-like rush.

Catmint has a long history of human use. It was traditionally employed to treat fevers and smallpox, and made into a tea with a lemony, minty flavour to treat coughs and colds, and to break up phlegm. It was consumed until the middle of the 16th century, when it was replaced in Europe by the 'true' tea plant (*Camellia sinensis*; p. 190). The remains of catmint have been found in old monastic gardens and the grounds of old hospitals. The cemetery of St Mary's Spital in east London is one of the most valuable archaeological sources for research into medieval English medicine. Catmint, borage (p. 58), opium (p. 82) and hyssop (p. 160) have been found during archaeological digs on the site, and are thought to have been combined to make a pain-relieving syrup. Catmint is strongly associated with sedation, and in the Middle Ages, if children were restless in sleep, small bags of catmint were hung above their beds or around their necks.

Strangely enough, the root of catmint has the opposite effect. Celtic warriors consumed it to ready them for war, and the 17th-century physician Thomas Sydenham reported: 'If the root be chewed it will make the most quiet person fierce and quarrelsome.' When the plant was introduced to North America in the early 19th century, hangmen took it before an execution, earning it the name 'hangman's root'.

The plant has been largely discarded for medicinal use owing to its supposed hallucinogenic properties. An article published in the *Journal of the American Medical Association* in 1969 suggested that it had cannabis-like effects. It turned out, however, that this was because the authors had confused catmint and cannabis (p. 90). It resulted in catmint's infamous reputation as a hallucinogenic, a conclusion that other studies have continuously disproved.

European myrrh

SWEET CICELY *MYRRHIS ODORATA* (APIACEAE)
COMMON USES DRINK • CULINARY

Myrrhis odorata, sweet cicely (also known as myrrh and sweet chervil), is not to be confused with the myrrh that was given to the baby Jesus by one of the three Magi at his birth. That gift came from the resinous tree in the genus *Commiphora*. Sweet cicely is a herbaceous plant of the carrot family, with feathery, fern-like foliage and umbrella-shaped flowers on hollow stems 1m (3ft) high. It flowers very early in the year, and its height helps it to stand out among other early-flowering plants in the competition for pollinating insects. It also has a long taproot, anchoring it in the soil and storing sugars and carbohydrates, allowing it to grow as a perennial in various climates and conditions.

Some believe that sweet cicely was introduced to Britain in 600 CE, while others think it came from central and southern Europe later, during the Middle Ages. Regardless, it has long been thought of as a native species. In medieval times it was used as a strewing herb. The species name *odorata* refers to the plant's fragrance, an aniseed scent that helped to cut through the smell of animals at a time when many people lived alongside them, under the same roof. Seeds of sweet cicely were used to polish wooden furniture, and Nicholas Culpeper said that the boiled roots did 'much to please and warm old and cold stomachs oppressed with wind or phlegm'.

Indeed, sweet cicely is an underappreciated herb, very rarely recorded in the herb books of the present day, despite a rich history of use. It was planted in monastic gardens as a fine ornamental species, and its tolerance of cold made it particularly popular in Scandinavia. In Trondheim, Norway, it was recorded growing alongside the prized angelica (p. 66) in special angelica gardens. It has been used as a diuretic and is being trialled as a treatment for malaria.

The Carthusian monks of La Grande Chartreuse in France used sweet cicely to make their famous liqueur, which is comparable to Greek ouzo and Italian Galliano. Both the leaves and the seeds can be used to sweeten food, and the botanist Roy Vickery suggests that sweet cicely was used during World War II to supplement sugar rations. The leaves can be eaten raw in salads, or used in cooking to add flavour. The fruit was recorded as a substitute for anise and fennel (p. 18), having a similar taste. Despite strong competition from these similar-tasting herbs, sweet cicely is still common in monastic and botanic gardens.

Sweet cicely has been used as a wood polish and also produces a yellow dye.

The beauty of roses often overshadows their toughness: they are a strong
species, able to grow in challenging environments.

A fine rose

ROSE *ROSA GALLICA* (ROSACEAE)
COMMON USES TEA • PERFUME

Roses are iconic flowering plants, symbols of love and affection, and therefore gifts truly fit for loved ones. These long-lived woody plants grow in upright, climbing or trailing form, with thick stems that can reach great distances.

Rosa gallica, which is thought to have originated in Central Asia, is a tough plant that can survive cold to -25°C (-13°F) and grows in full sun or shade. It was one of the first roses to be introduced to Europe, probably during the Crusades beginning in the 11th century, and is the parent plant of many of the hybrid roses that are grown today. Because it is known to hybridize in the wild, the characteristics of the original species are unknown. Dutch breeders first started making selections of *R. gallica* in the 18th century. They were followed by the French, and by the mid-19th century, French rose-breeders had created and selected more than 2,000 different cultivars. They went on to develop hybrids with more than one flush of flowers, rather than the single annual flowering of the wild rose.

R. gallica is believed to have first been cultivated some 3,000 years ago for aromatic and cosmetic purposes. When harvested, the petals retain their fragrance for several days, and they were often pushed into pillows to impart a fresh, sweet scent, or placed on the beds of newlyweds. They have been used in baths since the time of the ancient Greeks, who believed they would calm irritated skin and rashes.

Roses were used frequently in Christian ceremonies. In Austrian churches during the Middle Ages, a freshly cut stem of *R. gallica* was placed on the table or floor during important rituals. Roses – among them *R. gallica* and *R.* × *centifolia* – as well as species of *Dianthus* (carnation), *Convallaria majalis* (lily of the valley) and *Iris germanica*, were left as offerings for the Virgin Mary. Rose tea was popular among European monks, being caffeine-free, filled with antioxidants, and very high in gallic acid, which has known anti-cancer, antimicrobial, anti-inflammatory and analgesic effects.

Roses are everywhere in modern life. People enjoy their beauty and scent in gardens, and rose bouquets are given on Valentine's Day. The fragrance is used in perfume, cream and oil, and rose water – distilled from the petals – is used as a natural perfume by people who are sensitive to chemical scents. Rose water can soothe skin irritation and inflammation, heal cuts and burns, and alleviate sore throats. Rosebud tea is very popular throughout Central Asia, and *R. gallica* specifically is used in traditional Chinese medicine to treat menstrual pain.

Calming cosmetics

LAVENDER *LAVANDULA ANGUSTIFOLIA* (LAMIACEAE)
COMMON USES DISINFECTANT • PERFUME • CREAM

Lavender is a small aromatic shrub no more than about 1.5m (5ft) tall. Native to an area reaching from northeastern Spain to Italy, it grows in sunny conditions in alkaline soils. It is an evergreen plant, meaning that it retains its leaves all year round. The species name *angustifolia* is a descriptive one, meaning 'narrow leaf'. Lavender also produces purple flowers similar to those of other herbs in its family. In a dry, sunny situation it will flower for the whole summer.

Ancient civilizations used lavender in several ways. An old perfume factory was discovered in Cyprus in 2003, and the plant material found there – mostly lavender – was dated to about 2000 BCE. In ancient Egypt, the herb was part of the mummification process. Linen was soaked in a mixture of lavender oil and a simple cement, then wrapped around the bodies, which were left in the sun to dry out and harden.

In ancient Greece and Persia, lavender was used therapeutically, most commonly being added to baths and laundry for its scent and antiseptic qualities. The name 'lavender' probably comes from the Latin *lavare* (to wash). The Romans prized the plant and used the essential oil in hot baths to alleviate the stress and anxiety of battle, and the weariness of long marches. It was also used as a disinfectant, either applied directly to wounds or added to dressings. The Romans are largely responsible for the spread of the plant, having taken it with them to the outposts they established.

In the Middle Ages, lavender was placed in linen bags and tucked underneath pillows for its soporific effect, and also was used as a strewing herb. Its antiseptic qualities helped to reduce the spread of disease and sickness at a time of poor personal hygiene. Hildegard of Bingen in the 12th century recommended using lavender to get rid of lice, and also suggested that the strong scent could ward off evil spirits.

Lavender is a very important monastic plant. The garden of the abbey of Notre-Dame de Sénanque in Provence, France, can be visited today. It was founded in 1148 by 12 Cistercian monks, who created a garden to raise vegetables, obtained water from a nearby river, and operated a bakery. Over the next 60 years they built an abbey to live in. Unfortunately, it fell into disrepair owing to plague, war and religious attacks, and was abandoned. In 1988 Cistercian monks moved back into the abbey and restored this historic monument. Today, large fields of lavender and olive trees (p. 206) are beautifully set against the backdrop of the abbey, where the monks produce subtly fragranced lavender honey.

In the 16th century John Gerard recommended that a powder made from dried lavender, nutmeg, cinnamon and cloves (p. 213) be mixed with lavender water and drunk to treat heart problems, epilepsy and headaches. John Parkinson, the apothecary to King James I in the 17th century, wrote that lavender could be used to scent clothes, gloves, leather and linen, and that

In countries surrounding the Mediterranean, sprigs of lavender were at one time
woven into hats in the hope of preventing headaches caused by too much sun.

Those who suffer from stress, or struggle to fall asleep, may find it helpful
to use a lavender-scented pillow, or a pillow spray containing lavender.

the seeds could treat intestinal worms and headaches. Nicholas Culpeper echoed Gerard's and Parkinson's remedies, and added that a tincture made from lavender, cinnamon, fennel (p. 18) and asparagus root could ease headaches and toothache.

Lavender is also known as a treatment for burns. When the French chemist René-Maurice Gattefossé suffered chemical burns in 1910 after an explosion, he developed gas gangrene on his hands. After he washed them gently with lavender essential oil, the spread of the gangrene slowed and his healing escalated. He went on to work as a doctor during World War I, treating injured French soldiers with lavender oil, either applied directly to the wound using sphagnum moss or added to a water bath to clean it. Lavender is still used as an antibacterial agent in hospitals, alongside rosemary (p. 30), cinnamon, nutmeg and red sandalwood.

This strongly scented herb is cultivated extensively in Provence for perfume. It is also popular in local cuisine, adding its strong aroma to meat, sauces and desserts, such as ice cream and sorbet. Lavender tea, a calming infusion that aids sleep and reduces anxiety and nervousness, is also well liked. However, modern herbals and aromatherapy texts often identify *L. angustifolia* as the species used in tea, without providing more detail. It should be noted that some selections of *L. angustifolia* can have a strong spasmogenic (spasm-inducing) action, owing to high levels of camphor, so low-camphor selections are used in tea and for cooking. Other species, among them French lavender, *L. stoechas*, are used only in the cosmetic industry, since they contain too much camphor to be safe for consumption.

Lavender is an integral part of the global cosmetic industry, being used in a wide range of products from soaps, creams and shampoos to scented pillows and even veterinary products. Various studies into its essential oils have identified numerous health benefits. Lavender stimulates collagen, strengthening the skin and improving overall skin health; this is why it is often labelled as anti-ageing. The essential oils also contain anti-inflammatories, such as linalool and linalyl acetate, chemicals that reduce swelling and relieve itchy skin and joint and muscle pain. A study carried out in 2012 found that lavender essential oil significantly decreases blood pressure and heart rate.

Church pick-me-up

SOUTHERNWOOD *ARTEMISIA ABROTANUM* (ASTERACEAE)
COMMON USES DRINK • TRADITIONAL MEDICINE • BOUQUETS

Southernwood has adapted well to dry conditions and is distributed across Spain and Italy. This small bush with light, feathery foliage belongs to the same genus as wormwood (p. 74). It has a powerful scent, which varies according to the cultivar and origin, from citrussy to camphoric. There is even a cultivar, 'Cola', that has been bred specifically to smell like Coca-Cola.

Southernwood is a traditional medicinal herb in Europe, administered by the ancient Greeks and Romans to those with chronic respiratory ailments, such as tuberculosis and other lung diseases. It was transported along trade routes to the Middle East, where it was used to treat ulcers and skin diseases and was believed to ward off evil.

During the Middle Ages, southernwood spread further through Europe. Its strong association with monastic gardens gave it a reputation for being a church herb. People would keep a small twig of southernwood in their breast pocket and smell it if they found themselves growing tired during church services. Benedictine monks in Germany used it as a strewing herb, and the dried leaves were kept in wardrobes and scattered on valuable clothes to prevent them from being eaten by moths. Southernwood was also believed to be an aphrodisiac, as some of its traditional English common names attest: 'lad's love', 'maiden's ruin' and 'maid's passion'.

In the 10th century the Germanic peoples of central Europe and Scandinavia developed other uses for the herb, including the treatment of breathing problems, wounds and diseases of the organs. Hildegard of Bingen recommended southernwood as a poultice for wounds, and the herb was used (alongside rue; p. 69) by prison guards who believed that it would protect them from the contagious diseases of prisoners.

Southernwood was of particular significance in Sweden. It was the most commonly used herb in church flower arrangements, and was cultivated across the country, except Lapland. In Sweden's cold climate, the flowers do not open fully, so it is normally grown as a foliage plant. Swedish medieval literature records that it was once used to treat sleep-talking and 'female diseases', and was also included in sheep and horse medication. There it was called *abrot*, a word that Carolus Linnaeus incorporated into the Latin species name, *abrotanum*.

Southernwood is now found most often in homeopathic medicine. In 2004–5 the Charité University Medical Centre in Berlin conducted a two-year trial prescribing both herbal remedies and modern approved medicines containing southernwood. The results showed that southernwood was particularly helpful in reducing intestinal inflammation. The herb's documented antibacterial and antimicrobial qualities have led to its inclusion in cleaning products. The leaves are used in drinks, among them artemisia tea, absinthe and vermouth, and as a flavouring for meat and dairy products.

In Sweden women would include southernwood in the obligatory
bouquet they each carried when attending church services.

Cloves; see p. 213.

EXOTIC HERBS

Exotic Herbs

People have traded herbs since the earliest times. The ancient Egyptians and Greeks spent fortunes to transport herbs and spices from Asia, and in turn, European herbs and spices were sent to India and China. This fed the curiosity of both regions concerning non-native or 'exotic' plants.

An exotic plant is a species that is not native to a particular area, but has been introduced by humans, and is able to survive and reproduce there. For example, a European herb is considered to be an exotic in New Zealand because it could not be growing there without human intervention. Over the centuries, herbs have been introduced to new regions for several reasons – as medicinal, culinary or ornamental plants – or accidentally. The line between an exotic species and an invasive one can be thin. Some species are too well suited to their new environment, and begin to damage the local ecosystem and outcompete native plants. Some of the herbs mentioned in previous chapters, such as St John's wort and milk thistle (pp. 103, 111), are now recorded as weeds for this reason.

More positively, the global trade of plants has made the planet seem smaller. The domestication of plants is a universal human desire, and groups across the world have come together to trade and share knowledge about the native plants of their regions. The understanding of herbs and their uses has grown, and cultures have merged. The British are strongly associated with drinking tea (p. 190), for example, despite it coming from another continent; similarly cloves (p. 213), from a tropical tree, are now a treasured spice during the cold winters of northern Europe.

The trade route known as the Silk Road, established in the 2nd century BCE, is one of the most historically significant creations of global trade. Stretching for 8,000km (5,000 miles), it was – despite its name – not a single road but a network of routes across Asia. If we use today's borders, its path can be broadly described as beginning in Istanbul, and leading through Syria, Iran, Iraq and Central Asia to Xinjiang in western China. This storied route connected two continents. Before it was established, the Roman and Chinese empires knew little of each other beyond whispers and a few herbs.

Although trade was successful, some people wanted more. Europeans desired greater power over the trade in herbs, and grew tired of relying on imports. Explorers were sent in search of new plants. The British landed in India and colonized it, along with other countries in Africa and Oceania. The Dutch established strongholds in Indonesia and South Africa. The Spanish colonized the Americas and the Portuguese took strategic points along sea trade routes. These colonies became important hubs in the international trade of exotic herbs, and essentially monopolized the market at the expense of Indigenous people. This had devastating consequences for local communities and habitats, and came at a great cost to human life.

The plants described in this chapter illustrate the exchange of herbs from ancient civilizations through the period of discovery, describing their part in wars over monopoly and empire, their status in their adopted homes, and their availability in the present day. Exotic herbs and other food plants, such as cloves, star anise (p. 198), black pepper (p. 210), bananas (p. 214), cumin (p. 201) and olives (p. 206), are known around the world. Their availability in supermarkets is taken for granted, but few know the challenging journeys they have made to become so widely abundant. Alongside these iconic herbs are lesser-known species: sumac and lemongrass (pp. 202, 205). Sumac was the ancient Greek equivalent to lemon, with a tangy flavour that was valuable in cooking; while it is very popular in the Middle East, the rest of the world is less well acquainted with it. Similarly, many will have eaten the East Asian plant lemongrass, but most do not regularly cook with it.

The conqueror

TEA *CAMELLIA SINENSIS* **(THEACEAE)**
COMMON USES DRINK • ORNAMENTAL

Tea is an essential part of many people's lives. Some have an energizing 'cuppa' in the morning, while others drink it late in the evening while preparing for bed. Some drink a cup or three – or five or ten – daily. Some drink it as it comes, while others add a range of ingredients. Milk and sugar are common in Britain, lemon in Russia and mint in northern Africa. Tea is frequently served iced in Sri Lanka and the south of the United States.

The best-loved type of tea – called a world-conquering drink by the British anthropologist Alan Macfarlane – is made from the dried leaves of *Camellia sinensis*, a plant from East Asia. As with many traded herbs, tea's precise origin is unclear, but it is believed to come from a small distribution in China and India. The genus name was originally *Thea*, and the species were moved to the genus *Camellia* in 1818 by the English botanist Robert Sweet. *Camellia* belongs to a family that contains such fine ornamental genera as *Franklinia*, *Polyspora* and *Stewartia*.

The tea plant is an evergreen shrub with attractive flowers, well known for its unique fragrance and for needing acid soil. Some species can tolerate sun and, indeed, require its warmth to produce flower buds for the early spring or autumn, while others need shadier environments. It is important to protect the plant from drying winds, which will damage the leaves and flowers.

The origin of tea as a drink is also a matter of speculation. According to Chinese legend, in the third millennium BCE the mythological emperor Shennong was boiling water when tea leaves blew into the pot. He tasted the liquid and was impressed by the flavour, so he chose to share this new use with others. It is thought that tea plantations existed in China as early as 1100 BCE, and the documents suggest that tea was considered a regal drink, given to emperors. Evidence also indicates that tea leaves were used in religious offerings during the Western Zhou dynasty around 1000 BCE. The leaves were also eaten raw, finely chopped into salads or soups, and used as a bitter vegetable in cooking.

By the time of the Jin dynasty in the 3rd and 4th centuries CE, the drinking of tea had become a social occasion among the higher levels of Chinese society. The drink became more accessible to the lower classes through the expansion of tea plantations during the Tang dynasty (618–907 CE), and in 760 the philosopher Lu Yu published *Cha jin* (The Bible of Tea), the first tea-related monograph, listing the drink's uses, how to drink it, and its social status.

Tea spread into neighbouring countries along the Silk Road, and it was at this time that it became strongly linked with Buddhism. The growth of Buddhism and that of tea were intertwined owing to the frequent use of tea during religious ceremonies and to attain a state of tranquillity. Tea also has strong links to the Eastern philosophical movements Taoism and Confucianism.

The genus *Camellia* contains many notable species and cultivars, all
of them much-loved garden plants, in a range of colours and forms.

Tea does not cause the same caffeine 'crash' as coffee owing to its high levels
of antioxidants, which moderate the effect of the caffeine.

The Venetian explorer Marco Polo had collected tea in China in 1295, but it was not until the Dutch East India Company shipped the leaves from Java in the early 17th century that tea began to be established in the West. In Europe at that time, tea was a luxury product, but by the next century it was widespread throughout society. The Dutch held a monopoly over the import of tea to Europe before being overtaken by the English East India Company.

The word 'tea' comes from the Chinese Amoy dialect word *t'e*. When the plant was introduced commercially to Europe, the name was adapted by the Dutch to *thee*. In Mandarin, the word for tea is *ch'a*, in India it is *chai* and in Arabic *shai*. These words are still linked to tea, but are now commonly associated with spiced or herbal blends.

It was in the early 17th century that distinctions in the quality of different teas began to be made, using criteria based on origin, fermentation process and production. This led to a new range of tea products to satisfy the demand from tea-thirsty Europeans, among them Darjeeling and such Japanese green teas as Gyokuro. Teas are frequently blended, meaning that they are made from more than just the leaves of *C. sinensis*; fruit, perhaps, or flowers or spices are sometimes included. For example, the northern Indian *masala chai* contains such spices as cardamom (p. 155), cinnamon and cloves (p. 213), while Earl Grey tea is flavoured with the oil from bergamot (*Citrus × bergamia*).

Tea is well known for its range of health-improving compounds. It has been proved to suppress inflammation, cardiovascular disease and cancer. The consumption of tea was promoted during the COVID-19 pandemic because of its antiviral properties and the antioxidants it contains.

Demand has continued to grow since tea was discovered by those in the West. After water, tea is the most common drink in the world, being drunk by an estimated 3 billion people worldwide. While coffee tends to predominate in Europe and North America, tea surpasses it everywhere else. Both tea and coffee are produced in tropical or subtropical climates, almost entirely in developing countries. Yet only 29 per cent of coffee is consumed in developing countries, as opposed to 75 per cent of tea.

The global tea market was valued at around $200 billion in 2020, with an expected rise to $318 billion by 2025. Climate change, labour shortages and insufficient arable land currently restrict the expansion of tea production. Therefore, while high demand looks set to stay, we may see a rise in the cost of our beloved tea in the years to come.

Sweet and spicy root

GINGER *ZINGIBER OFFICINALE* **(ZINGIBERACEAE)**
COMMON USES COLDS • ASIAN CUISINE • DRINK

Ginger is a native of tropical Asia, India and southern China, and is widely grown there, as well as in Central America, Africa and Australia, as a commercial plant. The genus name *Zingiber* comes from the Indian term *inchi* (root). The plant is indeed grown mostly for its root, although the market for the candied stem is also strong. Ginger can be found growing naturally on tropical, damp forest floors, where it spreads by extending the thick underground rhizomes so familiar as an ingredient.

The history of its use in the West is long, closely associated with the East Indies spice trade of the 16th century, when demand for the dried root became very great. It was an expensive commodity, and rather than dealing with traders who brought it from the East, European explorers and entrepreneurs looked for sea routes to get directly to the crop.

Medicinally, ginger has been used for centuries to treat ailments, including colds, and for its anti-inflammatory properties. Ginger products are also used to settle the stomach and treat nausea, even quelling the effects of travel sickness.

Little surprise, then, that there is so much commercial production of this plant across the globe. Every region has cultivars bred specifically for rhizome size, keeping quality and robustness, to ensure competitiveness in the market. The resources it takes to produce the crop are worrying in the long term, however. Ginger is a labour-intensive crop that needs fertile ground and plentiful water. The length of time from planting to harvest varies according to the variety and the market for which it is being grown. Stems cut for confectionery are cropped in growth after five or six months, whereas the rhizomes are harvested after 6–18 months. Ground that is continuously used and cultivated has to be improved and fertilized for recropping. This intensive use makes the plants vulnerable to residual soil-borne problems, such as fungus and nematode damage.

Ginger is used in huge quantities all over the world, in such products as ginger snaps (cookies), tea, candied ginger, stir-fry sauces, curries, beer, ginger ale and wine. This familiarity has caused it to be absorbed into the English language, too. The use of 'gingerly' to describe caution alludes to the care with which such a strong spice would be added to a dish, while to 'ginger someone up' to encourage sprightly energy refers to the warming, stimulating effect of the spice. Both terms are used without thought of the knobbly root in the salad drawer. The same is true of a corruption of the botanical name, with the term 'zingy' being used to describe taste, colour and even fashion sense.

Ginger has become the cook's go-to exotic spice, as commonly used as garlic and black pepper (pp. 120, 210). Pot up a root and it will produce a handsome, leafy house plant, but in centrally heated homes away from its tropical origins it is unlikely to be added to the list of 'grow your own'.

Ginger is a tall, leafy perennial with striking flowers, and
grows naturally on damp, tropical forest floors in Asia.

In South Africa, rooibos tea is enjoyed cold
with lemon and sugar during the warmer months.

South African tea

ROOIBOS *ASPALATHUS LINEARIS* **(FABACEAE)**
COMMON USE TEA

Rooibos is endemic to South Africa, growing naturally in a small area from the north of the Cape Peninsula to Betty's Bay in the south. It is now under threat in the wild, however, owing to uncontrolled harvesting, overgrazing and climate change. This dry-adapted shrub, with slim, needle-like leaves, has the yellow flowers typical of members of the pea family.

The Indigenous uses of rooibos are uncertain. To date, the only archaeological material that has been found is fragments of plants recovered from charcoal in the Diepkloof Rock Shelter near Clanwilliam in the Western Cape. These were confirmed as rooibos in 2013 by the senior scientist Caroline Cartwright of the British Museum in London, and have been dated using electron microscopy to the Middle Stone Age (between 71,000 and 59,500 years ago).

It is not clear whether rooibos was used by early hunter-gatherers, and, if so, for what purpose. While it is often reported to be a traditional drink of the Khoekhoen and San people of South Africa, there are no recorded precolonial names for it: only words in English, Dutch and Afrikaans. This suggests that if it was used, it was probably only in a minor way. There is no credible evidence regarding Indigenous uses, since much of the traditional medicines and knowledge of local people has been lost.

To make the tea, young branches are cut and finely chopped to release the flavour. They are then watered and left to 'sweat', giving the drink its earthy flavour and red

colour. The earliest confirmed recorded use of rooibos was by the Swedish botanist Carl Peter Thunberg in 1772. Thunberg went to South Africa for three years after Dutch colonization to compile a flora of the land and of the Indigenous Khoekhoen people. He took a great interest in the rooibos plant for its potential as a tea product much cheaper than that imported from China, and in the second half of the 19th century rooibos tea became popular in South Africa for that very reason. It also started to gain attention in other countries, such as the Netherlands and Britain, where it was initially used as a herbal remedy to treat indigestion and other stomach complaints, and for its calming effect, particularly in young children.

In 1904 the businessman Benjamin Ginsberg, a Russian immigrant to South Africa, became the first person to export rooibos. Commercial growing followed in the 1930s, since when rooibos has slowly become a household name. It is now internationally popular, exported to more than 37 countries, of which Germany, the Netherlands, the United Kingdom and Japan are among the major consumers. Its popularity as a caffeine-free alternative to 'regular' tea (*Camellia sinensis*; p. 190) has made it second only to that globally. According to the South African Rooibos Council, rooibos provides employment for more than 5,000 people in the country. Some 14,000 tonnes are produced every year, to make the equivalent of 5.6 billion cups of tea annually.

Eight-horned spice

STAR ANISE *ILLICIUM VERUM* (SCHISANDRACEAE)
COMMON USES ANTIVIRAL · SPICE · INSECTICIDE

While the plant itself may not be particularly familiar, the star-shaped seed pods of *Illicium verum* are widely recognizable from the domestic spice rack. Star anise is well known for its aniseed flavour and is a common ingredient in the warm drinks of cold climates, such as mulled wine. It is also used for cold drinks in warm climates, such as iced tea in Thailand. Perhaps its best-known use is as a component of Chinese five spice, with fennel seeds (p. 18), black pepper (p. 210), cloves (p. 213) and cinnamon.

I. verum is a small evergreen tree from southern China and northern Vietnam, with strongly scented leaves and aromatic bark. It was previously classified in its own family, Illiciaceae, but is now in a small family with two other genera, both of which consist of woody plants that contain high volumes of essential oils. In evolutionary terms, star anise was one of the first plants to have flowers. The genus name comes from the Latin *illicere* (to entice or allure), while *verum* means 'true' or 'genuine'. The Cantonese and Mandarin names – *bat gok* and *ba jiao* respectively – mean 'eight corners', referring to the eight points of the fruit.

Star anise has a long history in Chinese traditional medicine. It was recorded in the herbal *Bencao gangmu* (Compendium of Materia Medica; 1578) by the Chinese scholar Li Shizhen. It is sometimes listed as *ba jiao hui xiang*, 'eight-cornered fennel'. The plant is consumed in different ways. The fruit is ground into a powder and used in tea to treat sleeplessness, the leaves are used to treat pain, and the essential oils alleviate rheumatism.

The plant contains an array of chemical compounds, including relatively high concentrations of the phenolic compound trans-anethole. The sweet flavour and scent of this compound mean that it is frequently added to perfumes, cosmetics and confectionery. The essential oil, on the other hand, is used against pests, as a component of synthetic insecticides and fumigants.

Star anise is also high in the chemical shikimic acid, which is used in the pharmaceutical industry as a central ingredient in the production of the anti-flu drug oseltamivir. This medication was used during the swine flu (H1N1) pandemic of 2009, although doubt has been cast over its safety after reports of serious side effects. It is important to note that other species of star anise, such as *I. anisatum* and *I. lanceolatum*, are toxic and should not be consumed.

Star anise is a key component of the popular ingredient Chinese five spice.

Some people are allergic to cumin, and can suffer from contact dermatitis,
in which they quickly develop a rash that persists for some days.

Do you take cumin?

CUMIN *CUMINUM CYMINUM* (**APIACEAE**)
COMMON USES CULINARY • CURRENCY

Cumin is a traditional, commonly used spice that grows best in dry, sandy areas with warm summers. It originated in northern Africa and the Middle East, but is now cultivated in many hot countries in the northern hemisphere, such as Uzbekistan, Turkey, Syria, India and Mexico. India dominates the market, accounting for nearly 90 per cent of global production.

The word 'cumin' and the genus name *Cuminum* are believed to derive from the ancient Semitic language spoken in the Akkadian empire, after about 2300 BCE, and passed into Arabic and Hebrew as *kammun* and *kammon* respectively. The Iranian city of Kerman – where cumin may have grown naturally, or was farmed abundantly – is thought to have been named after the plant. Confusingly, there are two other plants that share its name, both of them called black cumin; *Nigella sativa*, with hints of onion and oregano, is an unrelated species but shares some culinary uses with cumin, while *Elwendia persica*, from the same family as true cumin, has a more earthy, smoky flavour.

C. cyminum is documented in the ancient Egyptian artefact known as the Berlin Papyrus as part of a cough medicine that also contained milk and honey. Cumin had a range of uses, from stomach problems, tongue infections and complaints of the ears and teeth to the preservation of bodies during mummification.

The spice is mentioned many times in the Bible as a tithe, a form of tax whereby 10 per cent of an individual's income was donated in God's name to the church. Tithes were commonly payable in the form of prized herbs and spices, and the money was used to develop and expand the church. In the Old Testament book of Isaiah, the spent flowers of cumin are also documented as being beaten with a stick to help release the seeds (28:27).

The ancient Greeks used cumin as a flavouring for food, similar to the way black pepper (p. 210) is used today. At that time, cumin was cheaper and easier to get hold of. There seems to have been an old Greek expression related to cumin and frugal spending: the 4th-century Roman emperor Julian referred to his 1st-century predecessor Antoninus Pius as a 'cumin splitter'.

In the Middle Ages cumin became symbolic of love and was carried during wedding ceremonies for good luck, while scattering its seeds was said to prevent a lover from wandering. It was also believed to have aphrodisiac properties. However, as new, exotic herbs began to be introduced, cumin declined in popularity.

Cumin has a range of uses in the present day. It is best known as a spice used to flavour meat, vegetables, stews and broths. It is a staple of Indian dishes, and a key component of the spice mix garam masala. Aside from its culinary popularity, cumin is increasingly used in cosmetics owing to its high concentration of vitamins C and E, which promote skin health and counter premature ageing.

Dyed red

SUMAC *RHUS CORIARIA* (ANACARDIACEAE)
COMMON USES FOOD COLOURING • CULINARY

Being adaptable and very easy to grow, sumac has become a weed in many different climates. A shrub or small tree with long, slender green leaves topped with dense clusters of rich red flowers, it can cause a severe allergic reaction if direct contact is made, and even if you are not allergic to it at first, repeated exposure can cause a more serious reaction over time. Reactions, which resemble the symptoms of nut allergy and range from rashes to anaphylactic shock, are attributed to the chemical compound urushiol, which causes contact dermatitis.

The English word 'sumac' resembles the Arabian *sommaq* (red), referring to the colour of the fruit or perhaps the glowing autumnal hues of its foliage. The Latin *refus* (also 'red') may have become the genus *Rhus* over time. In ancient Rome, the fruit of the sumac was pressed and the resulting oil added to olive oil and vinegar to make sauces and dressings. Today, sumac is sold as a deep red powder made from the dried fruit, and used as a spice in Middle Eastern, Mediterranean and North African dishes. Its high concentration of malic acid gives it a tangy flavour similar to that of green apples and Cotton Candy grapes. The spice is combined with chicken, pine nuts and onions in *musakhan*, a Palestinian dish.

The vivid red of sumac lends itself to use as a natural food colouring, and it is also mixed with metallic compounds to produce a vast range of natural dyes. Its high tannin content meant that it was traditionally used for treating and colouring animal skins. All parts of the plant can be used for this purpose, but the leaves contain the most tannins. However, cheap synthetic dyes and tanning agents have been widely substituted for sumac in the leather industry since the middle of the 20th century. Sumac sap has antifungal properties that make it valuable for use in lacquer, to prevent wooden furniture from rotting and achieve a shiny finish, but it has also been largely replaced by the sap of Asian species (among them *Toxicodendron vernicifluum* and *T. succedaneum*).

Sumac's medicinal properties have long been recognized. Dioscorides wrote about its diuretic and anti-flatulent benefits in the 1st century CE, and in the early 11th-century *Canon of Medicine* by Ibn Sīnā, sumac is mentioned in the treatment of stroke symptoms. In traditional Iranian medicine, it has been used as a preventative agent for heart disease.

Sumac is now known to contain bioactive compounds that have antiviral, antifungal and anti-inflammatory qualities, and it is used to treat a range of health problems, including headaches, arthritis and digestive complaints. Researchers at Al-Mustansiriya University in Baghdad have conducted tests into the plant's effectiveness against bacteria, such as E. coli, with positive results. Additionally, sumac has shown great potential in the treatment of diabetes, balancing blood glucose and improving insulin function.

Western cookbooks between the 13th and 15th centuries featured sumac, a fact that may come as a surprise to present-day readers. One popular dish was *sumāqiyya*, which Europeans called 'somacchia'.

The oil from lemongrass is an effective insect repellent, except when it comes to honeybees, which are attracted to the scent. Beekeepers use it to help them relocate hives.

Eastern flavour

LEMONGRASS *CYMBOPOGON CITRATUS* **(POACEAE)**
COMMON USES CULINARY • DRINK • COSMETIC

This member of the grass family comes from Sri Lanka and southern India. It grows in warm tropical to subtropical regions and has consequently expanded its range, being now cultivated in Central and South America, Africa and other parts of Southeast Asia. It forms an evergreen tufted grass 1m (3ft) tall, with stiff, aromatic leaves. The genus name *Cymbopogon* comes from the Greek *kymbe* (boat) and *pogon* (beard), a reference to the plant's boat-shaped spathe (flower sheath) and fluffy white flower head. Unsurprisingly, *citratus* comes from the plant's lemony smell. Indeed, its essential oils are used in perfumes and other cosmetics.

This fragrance and the zingy flavour that accompanies it have long been used in Asian cuisine, and have made lemongrass a key ingredient in Thai cooking, which is noted for its vibrant flavours. It is best used fresh, since dried lemongrass has a less intense flavour. After the tough, fibrous outer leaves are removed, there are two possible preparation methods. In the first, the top and bottom of the stalk are cut away, leaving a piece about 10cm (4in) long. Next, the base of the stalk is bent, crushed or lightly cut, to allow the flavour to be released into the dish. This method is traditionally used in soups, such as the hot-and-sour coconut-milk soup *tom kha gai*, and other popular dishes, including the chicken recipe *gai yang ta krai*. The second method involves chopping the lemongrass finely and adding it to stir-fries; this is also used for the Vietnamese chicken dish *gà xiên nu'ó'ng*.

Lemongrass has long been used in Chinese traditional medicine to treat colds, abdominal pain and headaches. In Southeast Asia, it is made into a herbal tea that is used as a sedative, diuretic and anti-inflammatory. In both Egypt and Brazil, a similar drink is made from hot water and old, dried lemongrass leaves as an antispasmodic. In Indonesia and Malaysia, the whole plant is boiled in water and consumed as an aphrodisiac to increase blood flow to the pelvic area. The 20th-century herbalist and botanist James Duke, meanwhile, recommended the consumption of up to four cups of lemongrass tea per day, for its antifungal benefits.

A study conducted in Nicaragua in 2004 found that local communities were treating asthma using a syrup called *zacate de limon*, made from eucalyptus, oregano (p. 37) and lemongrass. In Brazilian folk medicine, the plant was believed to reduce anxiety. Many studies have explored this possibility, but have been unable to establish a link between the consumption of lemongrass and any reduction in anxiety.

Lemongrass oil has become increasingly popular as a natural insecticide. Its primary aim is to repel insects, such as mosquitos, that are vectors for malaria and other dangerous diseases. The antimicrobial compounds lemongrass contains, such as citral, make it suitable for use in detergents.

A branch of peace

OLIVE *OLEA EUROPAEA* **(OLEACEAE)**
COMMON USES FOOD • OIL

The olive plant originally comes from the Mediterranean region, hence the species name *europaea*. Its fruit graces the dining tables of many Mediterranean countries, and its oil is added to almost every dish. Countries have built wealth and expanded their influence because of the sale of this ancient plant.

The word 'olive' derives from the proto-Greek *elaia*, which dates from the early third millennium BCE. Over the years it became *olea*, a name that is recorded on a clay tablet from the 13th century BCE. The English word 'oil' comes from same root. The olive has been an important crop for thousands of years, and its productivity led to it becoming symbolic of peace, prosperity and fertility.

Olea belongs to a family that also contains such genera as jasmine (*Jasminum*) and ash (*Fraxinus*). *O. europaea* is a widely distributed species, and as a result is extremely variable in habitat. This means that its shape, leaf, fruit size and oil yield are not consistent from population to population. Such variations have been selected for cultivation, and there are now as many as 300 cultivars and named forms. As with many herbs with long histories, it is suspected that the true species itself may in fact be a hybrid. Olive trees take 10–15 years to first produce fruit, and the offspring of that fruit often differ from their parent plants.

Crushed olive stones have been found at the prehistoric archaeological site of Kfar Samir on the southern coast of Israel, suggesting that olives were eaten as far back as 6,500 years ago. There is strong archaeological evidence that the plants have been cultivated since at least the Bronze Age, and documents dating from Syria in 2000 BCE show that olive oil cost five times as much as wine.

The olive plant had a significant role in the ancient cultures of Egypt and Greece. When the Egyptologist Howard Carter discovered the sarcophagus of Tutankhamun in 1922, he found an olive wreath with it. Mummies were crowned with olive garlands in honour of the sun god Ra. The use of olive oil in treating ear infections of a 'foul-smelling' nature were also recorded in the Ebers Papyrus.

The ancient Greek writer Homer referred to olive oil as 'liquid gold', and Hippocrates called it 'the great healer'. It is said that when Athena, goddess of peace, threw her spear into the earth, an olive tree sprang from the base. This led to the idea of the plant's nobility. Cuttings of Athena's olive tree were believed to have been taken by a few specially selected people and passed down the generations. Olive trees can live to 2,000 years old, and some specimens are alive today that are thought to be much older. The practice of passing down single trees through family lines occurs in present-day Palestine.

The act of harming an olive tree was taken seriously by the ancient Greeks, and was punishable by death. Yet olives were also a symbol of mercy; criminals held

Mentions of olives can be found throughout religious texts, and
Islamic prayer beds are made from the wood of the olive tree.

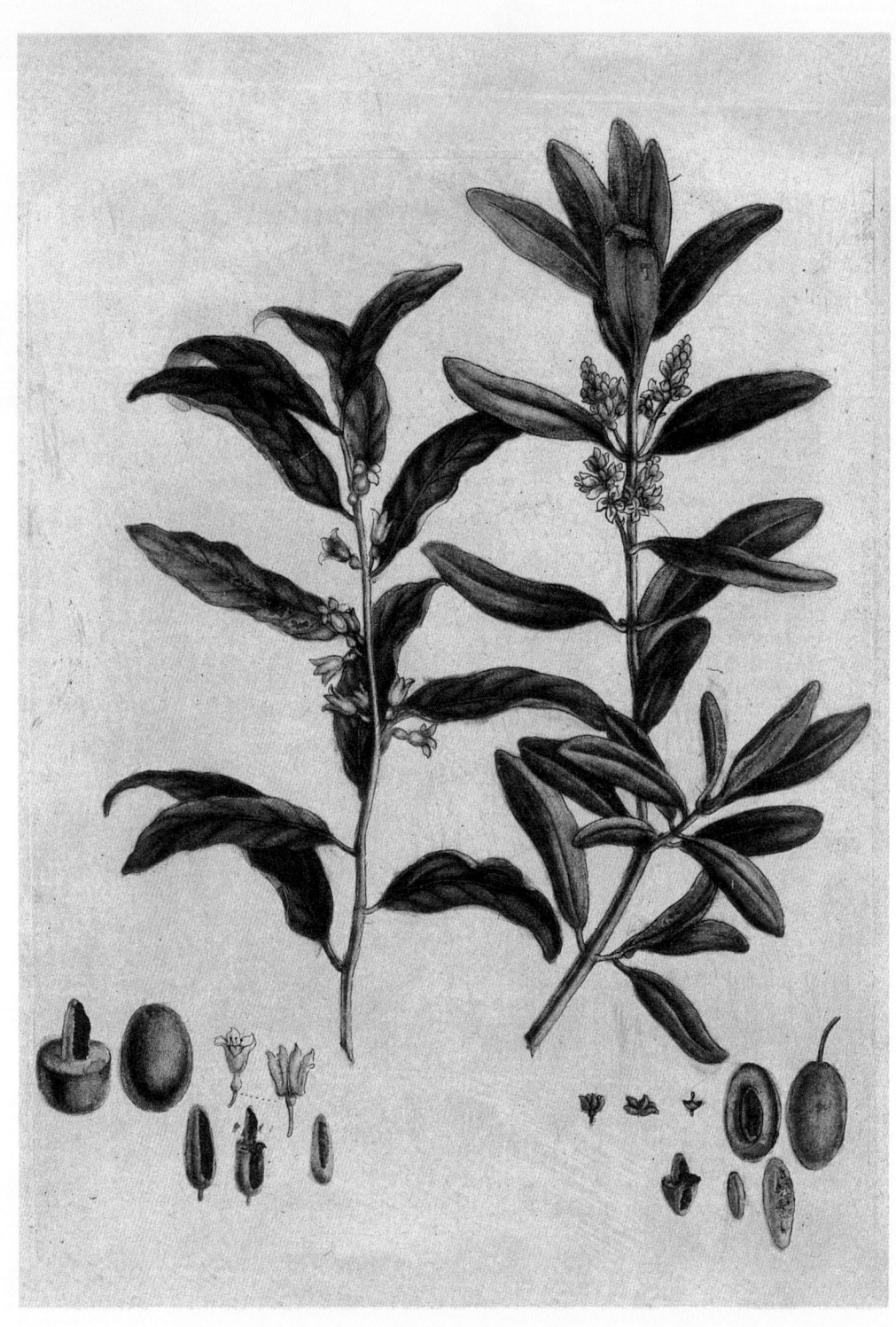

Olives were taken to South America by Spanish colonists, and are
perfectly adapted for growing in the climate of Chile, Peru and Argentina.

an olive branch to ask for leniency, and warriors raised them to indicate surrender. Olive branches appear on the flag of the United Nations as a symbol of peace and longevity. Wreaths of olive were placed on the heads of winners during the original Olympic Games, a tradition that was revived in Athens when the first modern Olympic Games were held, in 1896.

The cultivation of the olive tree has since spread to climates similar to its native Mediterranean, and it is now cultivated around the world, in Australia, California, Chile and Argentina. The heart of production, however, is still the Mediterranean basin. More than 90 per cent of the world's cultivated olives come from there, and the production of olives now takes up 9.4 million hectares (23.2 million acres) per year in that part of the world alone. Of that, some 6.5 million hectares (16 million acres) are used for olive oil, and only 607,000 hectares (1.5 million acres) are used to produce olives for consumption.

While many people now consider olive oil to be a culinary essential, this purpose was mostly ignored for a long period. Instead, its primary use was as lamp fuel, until the 19th and 20th centuries, when better, cheaper light fuels (such as kerosene, which was discovered in the 1840s) were invented, and it became redundant.

The process of collecting olives and making olive oil has remained mostly unchanged for 3,000 years. To produce the highest-quality oil, the fruit must be harvested without breaking its skin. It is then washed, but only lightly, or the natural oils will be removed and the flavour diluted. The olives are crushed and pressed to extract the oil, which is then stored, traditionally in the classic *amphorae* (tall clay pots with two handles often made by the Athenian pottery industry) and now in vessels made of glass or plastic.

Olive oil is well known for its health benefits and nutritional value. It is documented for decreasing the risk of cardiovascular disease and gastrointestinal problems, a benefit that is ascribed to the antioxidant phenols in the fruit. It also contains healthy fatty acids. Additionally, many phytochemical studies have been conducted into various compounds in olives, such as apigenin, beta-carotene and olivine, that have been selected for therapeutic use. These also have a role in improving the health of skin and protecting it against strong sunlight.

Black gold

BLACK PEPPER *PIPER NIGRUM* (PIPERACEAE)
COMMON USES SPICE • CURRENCY

Black pepper comes from a family whose members are distributed throughout the tropics. Peppercorns grow on a type of climbing vine with oblong leaves ending in a drip tip, as on many tropical plants, that helps to remove water from the top of the leaves and reduce the growth of algae. Unusually, the flowers have no petals or sepals, but grow as spikes; the unripe berries are green, turning black once dried. The fruit of a pepper, which contains a single hard seed, is botanically called a drupe – the same as a peach or plum.

The common name 'pepper' comes from the Sanskrit word for the plant, *pippali. Piper nigrum* is not to be confused with the closely related long pepper (*P. longum*), which was more popular at first. There are many unrelated species with common names that include the word 'pepper', among them Ethiopian pepper (*Xylopia aethiopica*), Guinea pepper (*Clethra alnifolia*), Chinese pepper (*Zanthoxylum simulans*) and the chilli peppers (*Capsicum* spp.; p. 147). What all these 'peppers' share is the numbing heat associated with the 'true' pepper, *P. nigrum*. This sensation is thought to be caused by the chemical piperine.

The Assyrians and Babylonians traded black pepper between 3000 and 2000 BCE, and later Arab traders took it to Rome and Egypt. It quickly became one of Rome's most prized spices, and in Egypt it was recorded in the Ebers Papyrus.

In about 1000 CE Rajaraja I, a powerful ruler in the southern part of India, extended his empire into Indonesia and Malaysia. At this time, the production of black pepper increased dramatically, and subsequently strong trade routes were established between the Indonesian island of Java and China. Later, India started its own production of black pepper in the southwest of the country, along the Malabar Coast. The spice was highly valued, and earned the nickname 'black gold'.

When the Portuguese reached India in the 16th century, the pepper trade was booming, and they were keen to benefit from it. They captured the Malaysian city of Malacca – a pivotal point on the trade route – and effectively ran the black-pepper trade from then until the Dutch East India Company arrived in the 17th century. The European colonial powers would battle for control of the trade over the next 300 years, joined in about 1795 by the newly independent United States.

Black pepper has been used to cure toothache by the Aztecs and the Jivaroans of northern Peru and eastern Ecuador, although the two peoples seem to have discovered this use independently. However, it is known best for its global culinary use, and is the most widely traded spice in the world. In 2020 alone, Vietnam produced 270,192 tonnes of black pepper, making it the world's largest producer and exporter of the spice.

Peppercorn rent is a traditional – if archaic – term of payment in the United Kingdom. A notable example
is the University of Bath, which pays a single peppercorn per year to the city for a large piece of land.

Clove oil is commonly found in dentists' surgeries, and is used as a local anaesthetic.

Cloves of smoke

CLOVES *SYZYGIUM AROMATICUM* **(MYRTACEAE)**
COMMON USES SPICE • CIGARETTES

The word 'clove' comes from the Latin *clavus* (nail), since the shape was thought to resemble a fingernail. This reference can also be found in the Chinese name for cloves, *ting-hsiang* (nail-shaped perfume). Cloves themselves are the dried flower buds of a large, tropical evergreen tree with bright red blooms.

The people of China have known of the benefits of cloves for nearly 2,000 years. They were believed to have restorative qualities, and to act as a powerful disinfectant. A range of medicinal uses were recorded in the first millennium CE in Asia: cloves were chewed to mask bad breath, or made into a tonic to treat toothache; and the wood was burned as a fumigant. Cloves were mixed with other herbs to make remedies for ailments including eye problems, rheumatism, respiratory problems and digestive complaints.

Cloves were initially understood by consumers in the West to come from China and/or India. During the Abbasid caliphate (750–1258), Chinese and Indian spices passed through Baghdad and Samarra, among other cities, on their way to Europe. Ibn Sīnā recorded cloves in *The Canon of Medicine* (1025) under the traditional name *quaranful*, and described how the dried buds were used to treat stomach and liver problems.

It was not until Portuguese explorers first landed in Indonesia in the 16th century that the true origin of cloves became known in the West. The Portuguese and later Dutch colonists took over the so-called Spice Islands to regulate the trade of cloves and nutmeg (*Myristica fragrans*): two spices that were expensive and highly sought-after in Europe. The colonization of these islands resulted in an oppressive system of farming that involved the enslavement of local people. The Dutch maintained their monopoly on clove trading from the mid-17th century to the 19th, before Tanzania and Zanzibar (colonized by the British) started producing the plants on a larger scale, thus breaking the monopoly.

A major use of cloves today is in cigarettes. Indonesia uses almost half of the global production of cloves to make *kretek* cigarettes, which are one part cloves, two parts tobacco (p. 86). The onomatopoeic name mimics the crackle of burning cloves. Although these cigarettes were created by one Haji Djamhari in the late 19th century to alleviate his chest pain and shortness of breath, the chemical compound eugenol, found in clove smoke, has in fact been linked to several cases of the lung infection aspiration pneumonia.

In Western Europe and the United States, cloves are strongly associated with Christmas, and feature in classic mulled drinks. The connection originated in the Middle Ages, when the spice was used in midwinter pomanders – from the French *pommes d'ambre* (amber apples) – oranges studded with cloves, to ward off evil spirits. Today, cloves are enjoyed in a range of dishes, and are staples of Russian, Chinese and Indian cuisine.

The world's biggest herb

BANANA *MUSA ACUMINATA* (MUSACEAE)
COMMON USE CULINARY

Bananas are among the world's best-known fruit, but their source is widely misunderstood. First, it is a herbaceous plant, not a tree, hence its inclusion in this book. *Musa acuminata* is the largest evergreen perennial in the plant kingdom. It has a 'pseudo-stem' of tightly wrapped leaves, which gives the plant its strength. Unlike trees, banana plants contain no woody tissue, and die back each winter in cold climates, pushing out new leaves in the spring. Below ground, a huge corm sucks up nutrients and water, allowing them to grow to unexpected heights for non-woody plants.

The second misconception concerns the fruit. Bananas are, botanically, large berries, while some of the fruit we call berries are not; strawberries, for instance, have seeds on the outside, making each one technically a collection of berries. Other botanical berries, such as cucumbers, tomatoes and grapes, produce seeds from the ovary inside, yet are not commonly thought of as berries. The banana is a berry because its seeds are inside.

Musa belongs to a family with only three genera. Some believe the genus name comes from Antonius Musa, a physician to the Roman emperor Augustus at around the time of Christ. Others suggest that it comes from the Arabic *mauz*, which appeared in Ibn Sīnā's encyclopaedia *The Canon of Medicine* in 1025. The common name is thought to have originated in western Africa.

Bananas were first cultivated some 7,000 years ago in Southeast Asia. One of the earliest descriptions was written by the Chinese scholar Chi Han in the early 4th century CE. He describes three different kinds of banana, and maintains that the long, pointed *yang-chiao-chiao* (sheep's-horn banana) is the sweetest and tastiest.

M. acuminata (from the Malay Archipelago) and the hardier *M. balbisiana* (from southern China) are the parents of today's banana plant. Cultivated bananas are classified based on their genetic content, for example AAB (where A is for *acuminata* and B for *balbisiana*). Commercial production is dominated by two varieties: Gros Michel and Cavendish. Gros Michel was introduced globally after being 'discovered' in Jamaica in 1835. Cavendish originated in China and is named after William Cavendish, the 6th Duke of Devonshire, who grew it in England in the mid-19th century. Owing to the limited genetics in commercial monocultural farms (those growing a single type of crop), global banana production is at great risk of disease.

The use of *M. acuminata* in food is broadly split between two selections: bananas and plantains. Plantains are starchier and always cooked, whereas bananas are sweeter and generally eaten raw. The leaves, although not eaten, can also be used in cooking. Food is wrapped in them and their aroma permeates the food. Their high water content stops them from burning, and they can then be used as a plate.

Despite its appearance, the banana is not a tree but a large
herbaceous plant, most of which can be used in cooking.

Select Bibliography

Biljana Bauer Petrovska, 'Historical Review of Medicinal Plants' Usage', *Pharmacognosy Review*, VI/11 (January–June 2012), pp. 1–5, www.ncbi.nlm.nih.gov/pmc/articles/PMC3358962

Elizabeth Blackwell, *A Curious Herbal* [1737] (John Nourse, 1739)

R Campbell Thompson, *The Assyrian Herbal* (Luzac & Co., 1924)

Michael Castleman, *The Healing Herbs: The Ultimate Guide to the Curative Power of Nature's Medicines* (Bantam Books, 1995)

Denys J Charles, *Antioxidant Properties of Spices, Herbs and Other Sources* (Springer, 2013)

Philip A Clarke, *Aboriginal People and Their Plants* (Rosenberg Publishing, 2011)

Nicholas Culpeper, *Culpeper's Complete Herbal* [1653], ed. Steven Foster (Sterling Publishing Company, 2019)

Iain Davidson-Hunt, 'Ecological Ethnobotany: Stumbling Toward New Practices and Paradigms', *MASA Journal*, XVI/1 (Spring 2000), pp. 1–13

Neveen Helmy Abou El-Soud, 'Herbal Medicine in Ancient Egypt', *Journal of Medicinal Plant Research*, IV/2 (February 2010), pp. 82–6

Hazem S Elshafie and Ippolito Camele, 'An Overview of the Biological Effects of Some Mediterranean Essential Oils on Human Health', *BioMed Research International* (November 2017), www.pubmed.ncbi.nlm.nih.gov/29230418

Steven Foster, 'Herbs for Health: Medicine in the Herb Garden', *Mother Earth Living*, 1 April 2001, www.motherearthliving.com/health-and-wellness/herbs-for-health-medicine-chest-in-the-herb-garden

Charis M Galanakis (ed.), *Aromatic Herbs in Food: Bioactive Compounds, Processing and Applications* (Academic Press, 2021)

John Gerard, *The Herball, or, Generall Historie of Plantes* (Norton & Whitakers, 1633), www.rct.uk/collection/1057467/the-herball-or-generall-historie-of-plantes

Roberta Gilchrist, 'Spirit, Mind and Body: The Archaeology of Monastic Healing', in *Sacred Heritage: Monastic Archaeology, Identities, Beliefs* (Cambridge University Press, 2020), pp. 71–109

David Gledhill, *The Names of Plants*, 4th edn (Cambridge University Press, 2008)

Maud Grieve, *A Modern Herbal* [1931] (Stone Basin Books, 2015), www.botanical.com

The Herb Society of America blog, www.herbsocietyblog.wordpress.com

Fez Inkwright, *Folk Magic and Healing: An Unusual History of Everyday Plants* (Liminal 11, 2019)

Victor Kuete (ed.), *Medicinal Spices and Vegetables from Africa: Therapeutic Potential against Metabolic Inflammatory, Infectious and Systemic Diseases* (Academic Press, 2017)

U Kumar Prajapati and Narayan Das, *Argo's Dictionary of Medicinal Plants* (Agrobios, 2003)

Gary J Lockhart, 'Herbal Scraps' (1997), www.arthurleej.com/HerbalScraps.pdf

Martin R McGann and Robert D. Berghage, 'The Pennsylvania State University Medieval Garden: Using a Specialized Garden as an Alternative Teaching and Learning Environment', *HortTechnology*, xiv/1 (2004), pp. 155–60

Si-Yuan Pan et al., 'Historical Perspective of Traditional Indigenous Medical Practices: The Current Renaissance and Conservation of Herbal Resources', *Evidence-Based Complementary and Alternative Medicine*, April 2014, www.ncbi.nlm.nih.gov/pmc/articles/PMC4020364

Sir Ghillean Prance and Mark Nesbitt (eds), *The Cultural History of Plants* (Routledge, 2005)

Cassandra L Quave et al., 'Medical Ethnobotany in Europe: From Field Ethnography to a More Culturally Sensitive Evidence-Based CAM?' *Evidence-Based Complementary and Alternative Medicine*, 2012, www.ncbi.nlm.nih.gov/pmc/articles/PMC3413992

Hercules Sakkas and Chrissanthy Papadopoulou, 'Antimicrobial Activity of Basil, Oregano and Thyme Essential Oils', *Journal of Microbiology and Biotechnology*, xxvii/3 (March 2017), pp. 429–38

Londa Schiebinger, *Plants and Empire: Colonial Bioprospecting in the Atlantic World* (Harvard University Press, 2004)

Ülle Sillasoo, 'Plants in Late Medieval Festivals and Customs in Written and Pictorial Sources from Southern Central Europe', *Environmental Archaeology*, xiv/1 (2009), pp. 76–89

Monique Simmonds, Melanie-Jayne Howes and Jason Irving, *The Gardener's Companion to Medicinal Plants: An A–Z of Healing Plants and Home Remedies* (Royal Botanic Garden Kew, 2017)

Alexandra Solomou et al., 'Medicinal and Aromatic Plants in Greece and Their Future Prospects: A Review', *Agricultural Science*, iv/1 (June 2015), pp. 9–20

Jack E Staub, *75 Exceptional Herbs for Your Garden* (Gibbs M. Smith, 2008)

Amy Stewart, *Wicked Plants: The Weed that Killed Lincoln's Mother and Other Botanical Atrocities* (Algonquin Books, 2009)

Theophrastus, *Recherches sur les plantes* [*Historia plantarum*], ed. and trans. Suzanne Amigues, 5 vols (Les Belles Lettres, 1988–2006)

Michael Tierra, *The Way of Herbs* (Gallery Books, 1998)

Roy Vickery, *Vickery's Folk Flora: An A–Z of the Folklore and Uses of British and Irish Plants* (Weidenfeld & Nicolson, 2019)

Harold Ward, *Herbal Manual: The Medicinal, Toilet, Culinary and Other Uses of 130 of the Most Commonly Used Herbs* (L N Fowler, 1967), www.swsbm.com/Ephemera/HerbalManual.pdf

Index

Page numbers in *italic* refer to illustrations; those in **bold** refer to main entries.

lemon verbena (*Aloysia citrodora*) 34, 50, 131, **136–7**
lemongrass (*Cymbopogon citratus*) 136, 189, **204–5**
leukaemia 103
Li Shizhen 140, 198
liver disorders/illnesses 14, 58, 65, 111, 152, 155, 163, 213
lovage (*Levisticum officinale*) **170–1**

M
malaria 74, 93, 107, 152, 163, 176, 205
Malaysia 205, 210
mandrake (*Mandragora officinarum*) 45, **46–9**, 62
meadowsweet (*Filipendula ulmaria*) 99, **108–9**
memory, improving 29, 30, 33, 58
menstrual pain 41, 66, 100, 107, 179
Mexico 33, 77, 86, 107, 119, 143, 147, 148
Middle East 22, 49
milk thistle (*Silybum marianum*) **110–11**
mint *see* peppermint
Morocco 34, 49
motherwort (*Leonurus cardiaca*) **166–7**
mouthwashes 34, 37, 41, 120, 132
muscle aches 37, 183

N
narcotics 54, 72–3, 82, 90
nausea 26, 152, 194

Netherlands, the 93, 155, 179, 189, 193, 197, 213
nicotine 86
North Africa 17, 34, 49
Norway 62, 66, 160, 176

O
oedema 14, 94
oils 90, 206, 209
  see also essential oils
olive (*Olea europaea*) 180, **206–9**
opium poppy (*Papaver somniferum*) 49, 62, 73, **82–5**, 175
oregano (*Origanum vulgare*) **36–7**, 205

P
painkillers 46, 49, 50, 57, 77, 81, 82, 85, 93, 107, 132
Pakistan 14, 93
Paraguay 131, 144
Parkinson, John 66, 100, 107, 180
Parkinson's disease 57, 81, 140
parsley (*Petroselinum crispum*) 12, **20–1**
peppermint (*Mentha × piperita*) **34–5**
perfume 29, 33, 108, 153, 155, 160, 179, 180, 183, 198, 205
Peru 119, 136, 210
pharmaceutical industry 34, 53, 85, 99, 100, 127, 130, 198
plagues 44, 45, 66, 69, 123
Pliny the Elder 26, 54, 58, 65, 69, 74, 81, 100, 103, 111, 120, 139, 163, 168
poisons 21, 46, 49, 54–7, 62,

65, 72–3, 77, 81, 89
  antidotes to 57, 62, 65, 89, 119, 132
pollinators, attracting 65, 94, 115, 120, 127, 132, 148, 167, 168, 176
Portugal 189, 210, 213
pregnancy 111, 164, 167
preservatives (food) 13, 14, 22, 25, 33, 37, 38, 120, 160

R
respiratory problems 33, 41, 50, 58, 112, 123, 139, 184, 213
rheumatism 29, 49, 81, 89, 93, 108, 119, 198, 213
Romania 22, 81, 167, 172
Romans 14, 18, 29, 30, 38, 49, 50, 58, 72, 98, 112, 120, 139, 155, 180, 202, 210
rooibos (*Aspalathus linearis*) **196–7**
rose (*Rosa gallica*) **178–9**
rosemary (*Salvia rosmarinus*) 8, 12, 13, **30–3**

S
sage (*Salvia officinalis*) 10, 13, **28–9**
St John's wort (*Hypericum perforatum*) 8, 99, **102–3**
salicylic acid 53, 108, 115
sciatica 41, 112, 119, 152
Scotland 55, 65, 66, 108, 116
sedatives 49, 62, 85, 93, 127, 136, 163, 164, 167, 168, 175, 205
self-heal (*Prunella vulgaris*) **104–5**
Silk Road 7, 72, 85, 139, 143, 188–9, 190

About the author

Connor Smith is head of one of Europe's largest rock gardens at the Utrecht University Botanic Gardens in the Netherlands. He has worked in Scotland, the United States, Germany, Italy and the Netherlands with a diverse range of plants. He gained his degree in horticulture with plantsmanship from the Royal Botanic Garden Edinburgh, has been published frequently by specialist plant societies, and lectures internationally.

Acknowledgements

I would like to thank my girlfriend, Marloes, who encouraged me to write and never complained about my absence from everyday life. I am grateful to my sister, Rachel, who supported me from the first word to the last. To Kevin Hobbs, for his guidance, mentorship and enthusiasm. To all family and friends who aided me. To Mark Fletcher, Rosanna Fairhead and John Round for helping me put together my first ever book.

Picture credits

First published in Great Britain in 2024 by

Greenfinch
An imprint of Quercus Editions Ltd
Carmelite House
50 Victoria Embankment
London EC4Y 0DZ

An Hachette UK company

A CIP catalogue record for this book is available from
the British Library.

Hardback ISBN 978-1-52943-053-0
Ebook ISBN 978-1-52943-054-7

10 9 8 7 6 5 4 3 2 1

Text by Connor Smith
Design by John Round Design
Printed and bound in China

Papers used by Greenfinch are from well-managed forests
and other responsible sources.